Leak Detection

Leak Detection

Technology and Implementation

Stuart Hamilton and Bambos Charalambous

IWA Publishing
London • New York

Published by

IWA Publishing
Alliance House
12 Caxton Street
London SW1H 0QS, UK
Telephone: +44 (0)20 7654 5500
Fax: +44 (0)20 7654 5555
Email: publications@iwap.co.uk
Web: www.iwapublishing.com

First published 2013

British Library Cataloguing in Publication Data
A CIP catalogue record for this book is available from the British Library

ISBN 9781780404707 (Paperback)
ISBN 9781780404714 (eBook)

FSC
www.fsc.org
MIX
Paper from responsible sources
FSC® C021018

Contents

About the Authors

Principal Author

Stuart Hamilton: shamilton@hydrotec.ltd.uk or stuart.hamilton@JD7.co.uk
Hydro Tec Ltd, The Barn, Thorpe Underwood, Northampton, NN6 9PA, UK.
JD7 Unit 8, Masons Business Park, Nottingham Rd, Derby, DE21 6YZ, UK.

Authors

Bambos Charalambous: bcharalambous@cytanet.com.cy
Managing Director, Hydrocontrol Ltd, Lemesos, Cyprus.

Mike Wrigglesworth: mike.wrigglesworth@soundprint.com
Pure Technologies Ltd., 300-705 11th Ave SW, Calgary, AB, Canada

Dale Hartley: hartley.d@sebakmt.com
Seba KMT, Dr-Herbert-Iann-Strasse 6, 96148 Baunach, Germany

Roger Ironmonger: roger.ironmonger@primayer.co.uk
Primayer Ltd., Primayer House, Parklands Business Park, Denmead, Hampshire PO7 6XP, UK

Mike Tennant: MTennant@hwm-water.com
Halma Water Management, Ty Coch House, Llantarnam Park Way, Cwmbran NP44 3AW, UK

Contributing Authors

Steve Tooms	stevetoomstmc@googlemail.com
Stuart Trow	stuarttrow@aol.com
Gosta Lange	lange@ultrac.se
Marc Bracken	marc@echologics.com
Andrew Chastain Howley	Chastain-HowleyA@BV.COM

Claude Gutermann	sales@gutermann.ch
Aristides Adamou	adamou@lwb.org.cy
Christodoulos Christodoulou	chris@lwb.org.cy
Sophie Kanellopoulou	sofia.kanellopoulou@live.com
Zheng Wu	Zheng.Wu@bentley.com
Tony Green	Antony.Green@advanticagroup.com

Acknowledgements

The authors would like to acknowledge the help and support of members of the IWA Water Loss Specialist Group.

Further acknowledgement

JD7 donated funds to publish this book in full colour.

JD7 is the world's leading company in CCTV, acoustic and condition assessment technologies.

Please visit www.jd7.co.uk

Chapter 1

Introduction

Ageing infrastructure and declining water resources are major concerns with a growing global population. Controlling water loss has therefore become a priority for water utilities around the world. In order to improve their efficiencies, water utilities need to apply good practice in leak detection.

The reasons for controlling leaks and reducing Non-Revenue Water have been well documented. Through the Water Loss Specialist Group and its Working Groups, the IWA has established several relevant guidelines, including the IWA Standard Water Balance and the Basic Management Strategies for Reducing Leakage.

To deal with losses in an effective manner, particularly from networks in water scarce areas, water utility managers are increasingly turning to technology to reduce costs, increase efficiency and improve reliability. Companies that continuously invest in technology and innovation should see a positive return on investment in terms of improving daily operations and collection and analysis of network data for decision making and forward planning.

The purpose of this document is to assist water utilities with the development and implementation of leak detection programs. Leak detection and repair is one of the components of controlling water loss. In addition to the techniques discussed within this document, water utilities should consider the other related Good Practices established by the IWA Water Loss Specialist Group.

Methodologies for achieving the best results to reduce water losses are continuously evolving. Water companies and equipment manufacturers are increasingly working together in an effort to stretch the boundaries of current knowledge. This is leading to some innovative technologies and new product development to complement current methodologies. This document reflects the situation at the time of publication.

Chapter 2

The technology matrices

The choice of a particular leak detection/location technique and technology depends on the operating conditions and construction material of the pipeline in question. To assist in making this determination, four different matrices have been developed.

(1) Mains fittings only – High Pressure
- For leakage detection on mains fittings only (no house connections) with pressures greater than 10 m head or 15 psi. Fittings are at a minimum distance of 200 m apart and maximum 500 m

(2) Mains fittings only – Low Pressure
- For leakage detection on mains fittings only (no house connections) with pressures less than 10 m head or 15 psi. Fittings are at a minimum distance of 200 m apart and maximum 500 m

(3) Domestic & Mains fittings – High Pressure
- For leakage detection on all property and mains fittings with pressures greater than 10 m head or 15 psi. Fittings are at a minimum distance of 10 m apart and maximum 50 m

(4) Domestic & Mains fittings – Low Pressure
- For leakage detection on all property and mains fittings with pressures less than 10 m head or 15 psi. Fittings are at a minimum distance of 10 m apart and maximum 50 m

The matrices consider the following pipeline materials:

- Metallic
  - Includes steel, ductile iron and other ferrous materials
- Concrete
  - Includes reinforced concrete, Pre-stressed Concrete Pipe (PCP)
- Asbestos Cement
- Glass-Reinforced Plastic (GRP)
- Polyvinyl chloride (PVC)
- Polyethylene
  - MDPE Medium Density Poly Ethylene
  - HDPE High Density Poly Ethylene

The technologies available are discussed in more detail later in this document. The equipment has been placed in the selected categories where it is reliably successful. The equipment may sometimes be successful in other categories but not reliably so.

Note that new equipment is continuously being developed: these matrices only take into account equipment that was available during the preparation of the matrices (up to December 2012).

2.1 MAIN PIPELINES ONLY – HIGH PRESSURE

This matrix is for leakage detection on mains fittings only (no house connections) with pressures greater than 10 m head or 15 psi. Fittings are at a minimum distance of 200 m apart and maximum 500 m.

Diameter	mm	*75*	*100*	*150*	*200*	*250*	*300*	*350*	*400*	*450*	*500*	*600*	*700*	*800*	*900*	*1000+*
	inches	*3*	*4*	*6*	*8*	*10*	*12*	*14*	*16*	*18*	*20*	*24*	*28*	*32*	*36*	*40+*
Material																
Metallic all		A,B, C,D, F,G	A,B, C,D, F,G	A,B, C,D, F,G	A,B, C,D, F,G	A,B, C,D, F,G	A,C, D,E, F,G	A,C, D,E, F,G	A,C, D, E	C,D, E	C,D,E	D,E	D,E	E	E	E
Concrete all		A,C,D	A,C,D	A,C,D	A,C,D	A,D	A,D,E	A,D,E	A,D,E	E	E	E	E	E	E	E
Asbestos Cement		A,C,D	A,C,D	A,C,D	A,C,D	A,D	A,D,E	A,D,E	A,D,E	E	E	E	E	E	E	E
GRP		A,D	A,D	A,D	A,D	A,D	A,D,E	A,D,E	A,D,E	E	E	E	E	E	E	E
PVC		A,D	A,D	A,D	A,D	A,D	A,D,E	A,D,E	A,D,E	E	E	E	E	E	E	E
Polyethylene all		A,D	A,D	A,D	A,D	A,D	A,D,E	A,D,E	A,D,E	E	E	E	E	E	E	E

Method A Gas Injection
Method B Traditional Techniques with Manual Listening Stick
Method C Non-Intrusive Acoustic Techniques that is Standard Correlator, Correlating Noise Loggers (Accelerometers)
Method D Intrusive Acoustic Techniques that is Standard Correlator or Correlating Noise Loggers (Hydrophones)
Method E Inline Inspection Techniques (Tethered & Free-swimming)
Method F Noise Loggers (Non-Correlating), Non-Intrusive Magnetic Connection
Method G Electronic Amplified Listening Ground Microphone

2.2 MAIN PIPELINES ONLY – LOW PRESSURE

This matrix is for leakage detection on mains fittings only (no house connections) with pressures less than 10 m head or 15 psi. Fittings are at a minimum distance of 200 m apart and maximum 500 m.

Diameter	mm	*75*	*100*	*150*	*200*	*250*	*300*	*350*	*400*	*450*	*500*	*600*	*700*	*800*	*900*	*1000+*
	Inches	*3*	*4*	*6*	*8*	*10*	*12*	*14*	*16*	*18*	*20*	*24*	*28*	*32*	*36*	*40+*
Material																
Metallic all		A,D	A,D	A,D	A,D	A,D	A,D,E	A,D,E	A,D,E	E	E	E	E	E	E	E
Concrete all		A,D	A,D	A,D	A,D	A,D	A,D,E	A,D,E	A,D,E	E	E	E	E	E	E	E
Asbestos Cement		A,D	A,D	A,D	A,D	A,D	A,D,E	A,D,E	A,D,E	E	E	E	E	E	E	E
GRP		A,D	A,D	A,D	A,D	A,D	A,D,E	A,D,E	A,D,E	E	E	E	E	E	E	E
PVC		A,D	A,D	A,D	A,D	A,D	A,D,E	A,D,E	A,D,E	E	E	E	E	E	E	E
Polyethylene all		A,D	A,D	A,D	A,D	A,D	A,D,E	A,D,E	A,D,E	E	E	E	E	E	E	E

Method A Gas Injection
Method B Traditional Techniques with Manual Listening Stick
Method C Non-Intrusive Acoustic Techniques i.e. Standard Correlator, Correlating Noise Loggers (Accelerometers)
Method D Intrusive Acoustic Techniques i.e. Standard Correlator or Correlating Noise Loggers (Hydrophones)
Method E Inline Inspection Techniques (Tethered & Free-swimming)
Method F Noise Loggers (Non-Correlating), Non-Intrusive Magnetic Connection
Method G Electronic Amplified Listening Ground Microphone

2.3 DOMESTIC & MAINS FITTINGS – HIGH PRESSURE

This matrix is for leakage detection on all property and mains fittings with pressures greater than 10 m head or 15 psi. Fittings are at a minimum distance of 10 m apart and maximum 50 m.

Diameter mm	75	100	150	200	250	300	350	400	450	500	600	700	800	900	1000+
Inches	3	4	6	8	10	12	14	16	18	20	24	28	32	36	40+
Material															
Metallic all	A,B,C, D,F,G	A,B,C, D,F,G	A,B,C, D,F,G	A,B,C, D,F,G	A,B,C, D,F,G	A,C,D, E,F,G	A,C,D, E,F,G	A,C,D, E,F,G	C,D,E, F,G	C,D,E, F,G	C,D, E	C,D, E	D,E	D,E	D,E
Concrete all	A,C, D,F,G	A,C, D,F,G	A,C, D,F,G	A,D	A,D	A,D,E	A,D,E	A,D,E	E	E	E	E	E	E	E
Asbestos Cement	A,C, D,F,G	A,C, D,F,G	A,C, D,F,G	A,C, D	A,C, D	A,D, E	A,D, E	A,D, E	E	E	E	E	E	E	E
GRP	A,C, D,F,G	A,C, D,F,G	A,C, D,F,G	A,C, D	A,C, D	A,D, E	A,D, E	A,D, E	E	E	E	E	E	E	E
PVC	A,C, D,F,G	A,C, D,F,G	A,C, D,F,G	A,D	A,D	A,D, E	A,D, E	A,D, E	E	E	E	E	E	E	E
Polyethylene all	A,C, D,F,G	A,C, D,F,G	A,C, D,F,G	A,D	A,D	A,D, E	A,D, E	A,D, E	E	E	E	E	E	E	E

Method A Gas Injection
Method B Traditional Techniques with Manual Listening Stick
Method C Non-Intrusive Acoustic Techniques that is Standard Correlator, Correlating Noise Loggers (Accelerometers)
Method D Intrusive Acoustic Techniques that is Standard Correlator or Correlating Noise Loggers (Hydrophones)
Method E Inline Inspection Techniques (Tethered & Free-swimming)
Method F Noise Loggers (Non-Correlating), Non-Intrusive Magnetic Connection
Method G Electronic Amplified Listening Ground Microphone

2.4 DOMESTIC & MAINS FITTINGS – LOW PRESSURE

This matrix is for leakage detection on all property and mains fittings with pressures less than 10 m head or 15 psi. Fittings are at a minimum distance of 10 m apart and maximum 50 m.

Diameter mm	75	100	150	200	250	300	350	400	450	500	600	700	800	900	1000+
inches	3	4	6	8	10	12	14	16	18	20	24	28	32	36	40+
Material															
Metallic all	A,C, D,F	A,C, D,F	A,C, D,F	A,C, D,F	A,D	A,D,E	A,D,E	A,D,E	E	E	E	E	E	E	E
Concrete all	A,D	A,D	A,D	A,D	A,D	A,D,E	A,D,E	A,D,E	E	E	E	E	E	E	E
Asbestos Cement	A,D	A,D	A,D	A,D	A,D	A,D,E	A,D,E	A,D,E	E	E	E	E	E	E	E
GRP	A,D	A,D	A,D	A,D	A,D	A,D,E	A,D,E	A,D,E	E	E	E	E	E	E	E
PVC	A,D	A,D	A,D	A,D	A,D	A,D,E	A,D,E	A,D,E	E	E	E	E	E	E	E
Polyethylene all	A,D	A,D	A,D	A,D	A,D	A,D,E	A,D,E	A,D,E	E	E	E	E	E	E	E

Method A Gas Injection
Method B Traditional Techniques with Manual Listening Stick
Method C Non-Intrusive Acoustic Techniques that is Standard Correlator, Correlating Noise Loggers (Accelerometers)
Method D Intrusive Acoustic Techniques that is Standard Correlator or Correlating Noise Loggers (Hydrophones)
Method E Inline Inspection Techniques (Tethered & Free-swimming)
Method F Noise Loggers (Non-Correlating), Non-Intrusive Magnetic Connection
Method G Electronic Amplified Listening Ground Microphone

Chapter 3
Acoustic principles

As many of the technologies currently used for leak detection involve acoustics, it is important to understand some basic principles of leaks, and some general physics involved. The noise characteristics of a leak have been used for many years to locate leaks – listening on valves, hydrants, stop taps, or at the ground surface above the line of the pipe.

3.1 HISTORY OF ACOUSTICS

Many scientists and researchers over the centuries have been experimenting with sound and acoustic theory in order to discover and formulate solutions relating to a number of practical problems. One of the first people to experiment with underwater acoustics was Leonardo Da Vinci in 1490 and documented his thoughts by discovering that if you are on a ship and bring it to a halt, by placing a long tube in the water you will be able to hear by placing your ear on the end of the tube ships that are far away from the you. Isaac Newton subsequently developed mathematical principles which dealt with sound. However, a major step in the history of acoustics was made by Charles Sturm, a French Mathematician and Daniel Colladon, a Swiss Physicist. Their experiment took place on Lake Geneva in 1826 where they measured the time difference between a flash of light and the sound of a submerged bell. The experiment was a success and the speed of sound measured was 1435 metres per second over a distance of 17.000 metres. This was the first time that a quantitative measurement was carried out and this sound speed value remains within a margin of acceptance of about 2%. Modern acoustic theory was established and documented by Lord Rayleigh in 1877.

Underwater acoustics became extremely important with the start of the World War I with anti-submarine listening systems being developed. A number of echolocation patents were granted in Europe and the United States of America with Reginald A. Fessenden's echo-ranger being patented in 1914. At the same period in France Paul Langevin and in Britain A. B. Wood and associates were carrying out similar pioneering work. Active ASDIC (from Anti-Submarine Detection Investigation Committee) and passive SONAR (SOund Navigation And Ranging) were developed during the war, enabling the first large scale deployments of submarines. Acoustic mines were also another great advancement in underwater acoustics.

The refraction of sound rays produced by temperature and salinity gradients in the ocean were first described in a scientific paper in 1919. The range predictions were experimentally validated by transmission loss measurements.

Applications of underwater acoustics developed during the next two decades after the First World War. In the 1920's, commercial developments included the fathometer, or depth sounder and natural materials were used for the transducers. By the 1930s sonar systems incorporating piezoelectric transducers made from synthetic materials were being used for passive listening systems and for active echo-ranging systems. These were used extensively during World War II by both submarines and anti-submarine vessels.

Advances in the theoretical and practical understanding of underwater acoustics have been aided largely in recent times by computer-based techniques. The methodology applied today to detect water leaks using leak noise correlators and noise loggers is based on the principles of underwater acoustics.

3.2 PROPAGATION

Water escaping through a leak creates a noise. The sound waves propagate along the pipe wall, fittings, surrounding ground and especially via the water inside the pipe. If the pipe wall were completely rigid, the sound would propagate with a velocity of approximately 1485 metres per second. However, the pipe material is always elastic to some degree. This elasticity causes attenuation of the pressure wave as it progresses down the pipeline.

The sound velocity in water pipes depends on the pipe material and the ratio between the diameter and wall thickness. For metallic pipes, the sound velocity slows down to about 1200 m/s, although the metal absorbs only a fraction of the sound energy and the sound still travels quite far. Plastic pipes are much more elastic, reducing the sound velocity to 300–600 m/s. Furthermore, the sound energy is absorbed more easily causing the sound waves to become weaker and weaker as they travel along the pipeline.

3.3 RESONANCE

Every pipe will exhibit a certain resonant frequency; if only longitudinal sound waves are considered (circumferential resonances will also appear but are of less importance). This resonant frequency is dependent upon the physical dimensions of the pipe and also upon the velocity of sound. It will therefore be particularly low for plastic materials, but also low for metal pipes of larger diameters. It can often be as low as around 10 Hz, which is well outside the perception range of the human ear (20–20 000 Hz in a healthy young person).

3.4 ATTENUATION

Higher frequencies are always attenuated more strongly with distance than lower frequencies. One example can be whales in the ocean, who communicate over enormous distances at subsonic frequencies. Another is distant thunder, which is only perceived as a low frequency rumble. However even low frequencies are eventually attenuated over long distances.

When a leak noise is attenuated enough, it will be masked by other noises such as traffic, and ambient noises from effects within the pipe (such as turbulence resulting from rough surfaces) until ultimately even an expert cannot discern the leak sound.

Low frequencies are less attenuated and travel farther before they drown in the ambient noise. The snag is that human hearing cannot respond to the lowest frequencies.

The attenuation of a sound wave is small at resonance. Below the resonant frequency, attenuation increases slightly with decreasing frequency, whereas above the resonant frequency attenuation will increase strongly with increasing frequency.

3.5 ACOUSTIC IMPEDANCE

Every material has certain acoustic impedance that is expressed as the product of its density and its speed of sound.

If a sound is travelling in a certain medium, for example water, and meets a medium with different acoustic impedance, for example air, a part of the sound wave will be reflected back.

If the differences in acoustic impedance are great, almost all of the sound will be reflected. If on the other hand both media have the same acoustic impedance, the sound will travel through the boundary with no reflection. In practice, the case lies between those extremes, and reflection will be partial.

The air/water boundary will reflect practically all sound, because both the density and the speed of sound of the two media are so different. The water/steel boundary will also reflect most of an impinging sound, since both the density and the speed of sound of steel are much greater than the corresponding quantities for water.

Chapter 4

Leak detection technologies

There are a vast number of techniques to detect where leakage is occurring in the network. Location accuracy depends on many factors, and the subsequent portions of this document provide further detail. Some techniques are able to approximate or localize the position of a leak while others can find exact locations. Often a tool-box approach is used, where multiple technologies are deployed.

4.1 METHOD A: GAS INJECTION METHOD

This method uses a gas detector to find the presence of a tracer gas that has been injected into a pipeline. While helium can be used, the most common tracer gas is hydrogen due to its lower cost and high performance.

Hydrogen is the lightest gas and has the lowest viscosity. This makes it easy to fill, evacuate and dissipate. Typically diluted 5% in nitrogen, the gas can be injected into buried and ducted cables, pipelines and also small diameter in-house heating pipes.

The gas injection method can be used to detect leaks in all pipe materials from 75 mm to 1000 mm in diameter. It can be used on pipes of greater diameter but for obvious reason a considerable amount of gas would be required. The pipeline can be empty of water or full, however with the pipeline full of water less gas is required to be used to find the leak.

To accurately locate the leaking gas which comes to the surface after leaving the leak in the pipe, the direction of the water flow must be known and the gas should be kept within the pipeline of where the leak is suspected. This requires the closure of any branches/off takes which may cause the gas to be diluted or transferred away from the pipeline in question. The mixing of the gas with water does not affect the water quality. This methodology can be used in all types of sealed tubes including cables and pipelines. The material has no effect on the gas injected.

4.2 METHOD B: MANUAL LISTENING STICK

The stethoscope or listening stick has an earpiece and is used to listen to leaks on fittings and to pinpoint the location of a leak. It is a widely used piece of equipment for many water utilities. The material of the listening stick can be metal, wooden or plastic. This technique is dependent on the ability of the engineer to hear the leak and uses no electronic equipment to enhance the sound.

This technique is best suited for use on metallic pipelines between 75 mm and 250 mm and with pressures above 10 m (15 psi). The material or pipe size does not prevent the listening stick from being able to pinpoint the leak from the surface, but what does affect this is the type of leak, ground backfill material, pressure of the water leaving the pipe, background noise and the ability of the engineer (Figure 4.1).

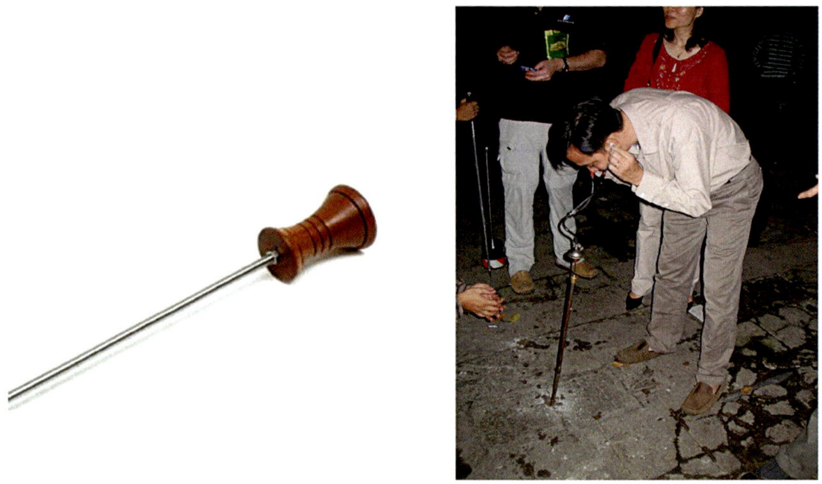

Figure 4.1 Manual Listening Sticks.

4.3 METHODS C AND D: LEAK NOISE CORRELATION

Leak noise correlation works by comparing the noise detected at two different points in the pipeline. Assuming consistent pipe material and diameter, the noise travels from the leak in both directions at a constant velocity, so that if the leak is equidistant between two sensors then these sensors will detect the noise at the same time. Conversely, if the leak is not equidistant then the sensors will detect the same noise at different times – this difference in arrival times is measured by the correlation process.

The following diagram illustrates this principle (Figure 4.2):

The sensors are located on valves A and B (convenient access points for underground pipes), and as shown the leak position is closer to A.

By the time an instance of noise from the leak has reached A, the same noise heading towards B has only travelled as far as point X. The distance from X to B causes a time delay (t) before the noise arrives at B. The correlation processing detects the delay (t) between the arrival of the noise at A and its arrival at B. If the velocity of sound is V and the distance between the loggers is D, then as the distance from X to $B = V * t$.

Then $D = (2 * L) + (V * t)$.

This equation may be rearranged to give L, the distance from the nearer logger to the leak site:

$$L = \frac{D - (V * t)}{2}$$

Correlation measures the time delay (t). The distance between the sensors must be determined by accurate measurement.

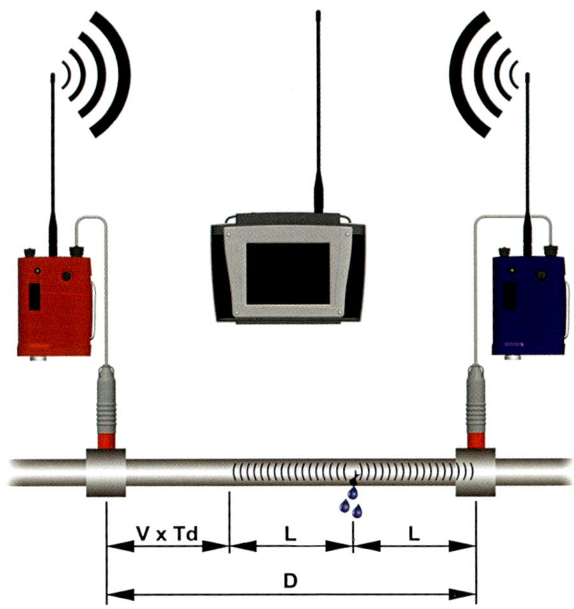

Figure 4.2 Principle of Correlation (*Source*: Halma Water Management).

The sound velocity depends upon pipe material, pipe diameter and, to a lesser extent, on surrounding soil. Often theoretical values of sound velocity are used and this is fine for a first approximation of the leak position. However, the velocity will vary due to many factors, and significantly so if a repair section of a different pipe material exists. Sound velocity must therefore be measured or, alternatively, multiple correlations carried out.

With all correlation techniques practitioners should be aware that any noise source can result in a correlation peak and all results should thus be treated as 'points of interest' until confirmed. Confirmation is usually done using a ground microphone.

It is important to note that the capability of correlators is dependent on the pressure and level of background noise within the network. Furthermore, correlation can become impossible, because it requires two monitoring points, one on each side of the leak and the attenuation often causes leak signals to disappear at one or both points.

Leak noise correlation requires a noise signal. There are two types of noise transducer in normal use: accelerometers and hydrophones. These have significant differences in their deployment and uses, and have hence been identified as two different methods in the sections below.

4.4 METHOD C: CORRELATION USING ACCELEROMETERS

To perform correlation using accelerometers, two sensors are deployed on pipe fittings, so no access to the water inside the pipe is required. The sensors are then positioned on either side of the suspect leak position. Accelerometers respond to acceleration and so tend to be more responsive to higher frequencies Accelerometers are most effective on metallic pipes and tend to be less effective with non-metallic pipes. This is due to both the rapid attenuation of high frequency signals in many non-metallic pipes and the impedance mismatch between the pipe material and the metal fittings normally used to site accelerometers. However their ease of deployment and low-cost make them an attractive choice when trying to localize leaks (Figure 4.3).

Figure 4.3 Example of Accelerometer (*Source*: Halma Water Management (left) and Primayer (right and bottom)).

4.5 METHOD D: CORRELATION USING HYDROPHONES

Leak location on plastic pipes and in large diameter pipes is difficult because they become more elastic as the pipe material gets softer, or the ratio of pipe wall thickness to overall diameter increases. This results in more rapid absorption of leak energy into the pipe material and surrounding soil, with resultant attenuation as distance from the leak increases.

Furthermore, higher frequency leak noise energy is absorbed much more rapidly than low frequency energy.

Accelerometers placed on the outside of the pipe detect the energy lost into the pipe wall. The water-borne wave is detected directly by hydrophones placed into the water at convenient fittings such as fire hydrants. Combined with advanced filtering technology, particularly at very low acoustic frequencies, hydrophones are beneficial in enabling leak location on plastic pipes and on larger diameter mains (Figure 4.4).

Figure 4.4 Example of Hydrophone (*Source*: Primayer (top) and Seba KMT (bottom)).

Hydrophone-based acoustic correlation techniques are beneficial in locating leaks in difficult situations. These situations are likely to occur in the following:

- Large diameter or trunk mains
- Plastic pipes
- Pipes with large distances between available pipe access points
- When high acoustic background noise exists, often due to traffic, and so on.

4.5.1 Technologies for leak noise correlation

4.5.1.1 Radio based correlator

Accelerometers or hydrophones are deployed on hydrants or similar fittings available with suitable pipe threads. They are positioned on either side of the suspected leak position and the noise created by the leak is detected by the accelerometer/hydrophones. This signal is transmitted via a radio signal to the correlation processing device (Figure 4.5).

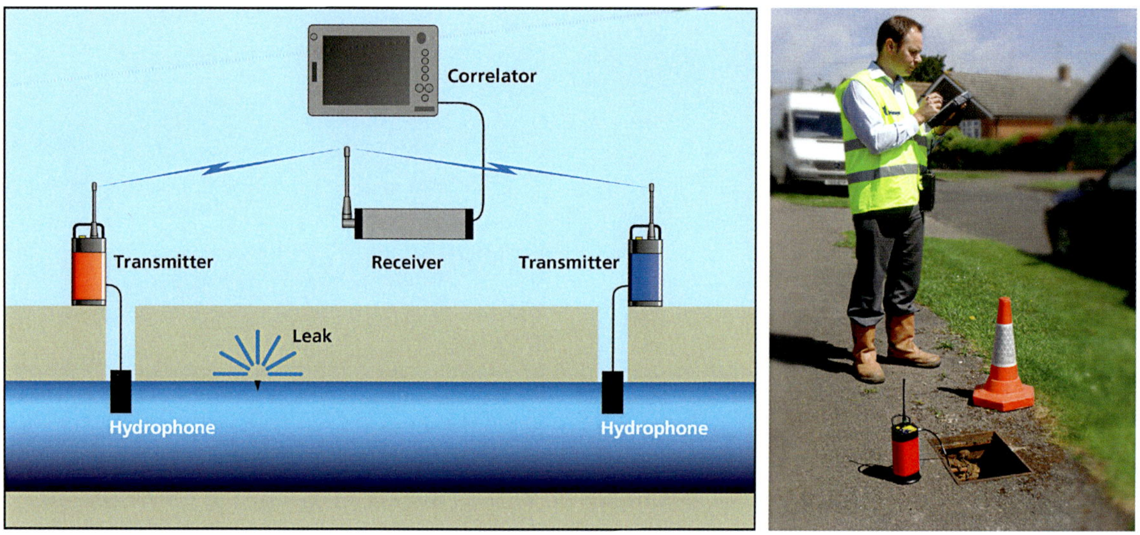

Figure 4.5 Two point radio-based system (*Source*: Primayer).

The system typically consists of:

- Correlator or laptop computer with advanced correlation software
- Accelerometer or Hydrophone sensor + pipe fittings (2 or 3)
- Radio transmitter (2 or 3)
- Radio receiver unit
- Vehicle antenna
- Headphones
- Battery charger
- Transport case

For optimum performance, when detecting low level leak noise, the complete correlation system design is optimised to reduce inherent electronic noise (always present in electronics). This means that radios will usually be digital or correlating loggers are used to drastically reduce the electronic components in the system.

Systems for these applications will also have a very low frequency response, capable of operation from a few Hertz upwards. With the processing power of modern computers a variety of advanced filtering techniques are now available to further enable a correlation result to be successful. These may be applied during on-site correlation processing and during post-processing (back in the office).

Correlation processing and presentation of results varies greatly between manufacturer's products. All products will require the entry of distance between sensors plus details of pipe material (to estimate the velocity of sound). Most will also allow entry of user defined material/velocity values.

Filters are a powerful facility to assist in obtaining a correlation peak on modern correlation systems. Filtering techniques vary but can include:

(a) ***Manual filtering***
   - These allow the operator to set filters based upon experience or what can be heard via headphones.

(b) ***Auto-filters***
   - These scan the received signals and automatically determine one, or more, filter band for optimum correlation.

(c) ***Coherence***
   - The Coherence Function is a computed value that gives a measure of the similarity between the two leak noises as a function of frequency. So the frequencies at which the coherence is high are likely to give a good correlation result. Coherence is not fool-proof, but can often lead to good results (Figure 4.6).

4.5.1.2 Advantages and disadvantages of radio-based correlators

Advantages

- Gives good results in many situations.
- Two point radio based correlators are generally fast to deploy and give rapid results whilst on-site.
- If a correlation result is not achieved then the opportunity exists for deployment at alternative pipe fittings.

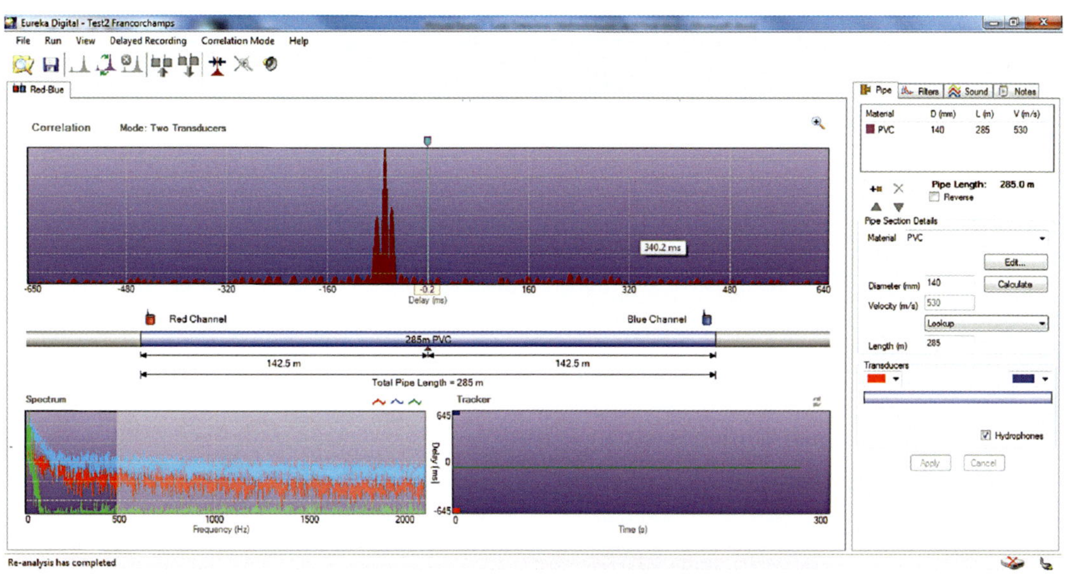

Figure 4.6 Display showing correlation function together with frequency spectrum, pipe model and pipe entry details. Result is on 140 mm PVC pipe over 285 metres (*Source*: Eureka Digital, version no. 1.8.1.0, Released 2010, Primayer Ltd, Denmead, UK).

- Digitally recorded sounds can be further processed off-site.
- Addition of third radio means that the velocity is measured and hence the leak position result calculated with much higher confidence.

Disadvantages

- Radio transmitters are sometimes difficult to deploy in busy urban streets when traffic disruption occurs.
- Will detect consumer use and non-water leak noises.
- Use of stored, theoretical, velocity values can give erroneous results.
- Limitations of radio range. Radio range can vary greatly depending upon proximity of local buildings and weather conditions. This only becomes a problem when working over long distances with limited access points.

4.5.1.3 Multi-point correlating loggers

Correlating loggers allow multi-point correlation and this technique can be used for combining leak detection and location. On large diameter trunk mains they may be placed strategically along the pipe. The example below shows eight correlating loggers on a PVC pipe network – 28 correlations are produced which can be sorted on 'confidence.' Two leaks are located, by many loggers (Figures 4.7 & 4.8).

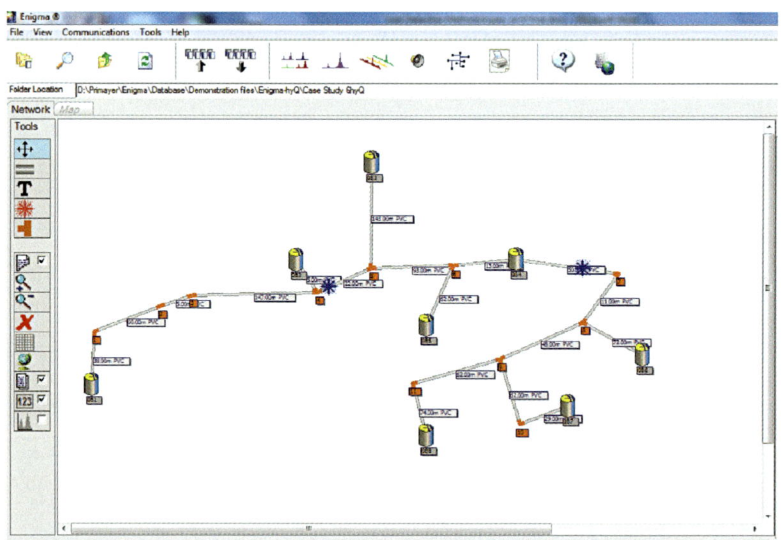

Figure 4.7 Correlating loggers on a PVC network (*Source*: Enigma, version no. 1.4.11.0, Released 2010, Primayer, Denmead, UK).

The system normally consists of:

- Laptop computer with advanced correlation software
- Correlating loggers
- Hydrophone sensors + pipe/hydrant fittings (Figure 4.9)
- Communications case

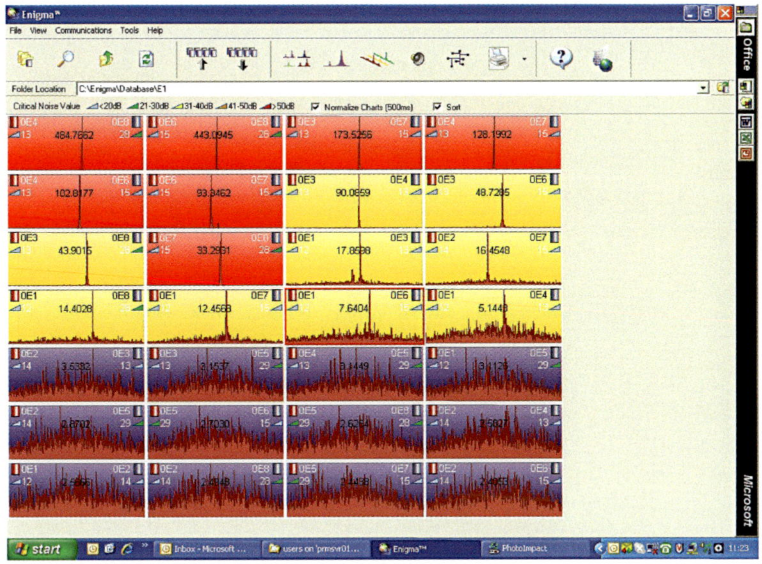

Figure 4.8 Correlating loggers on a pipe schematic together with resultant correlations (*Source*: Enigma, version no. 1.4.11.0, Released 2010, Primayer, Denmead, UK).

The entry of distance plus details of pipe material is required between all loggers. However, if carried out after the correlation 'run' then only those logger pairs with a correlation result require entry of these details. These are most easily entered on a map or pipe schematic.

As described above, filters are a powerful facility to assist in obtaining a correlation peak (Figure 4.10).

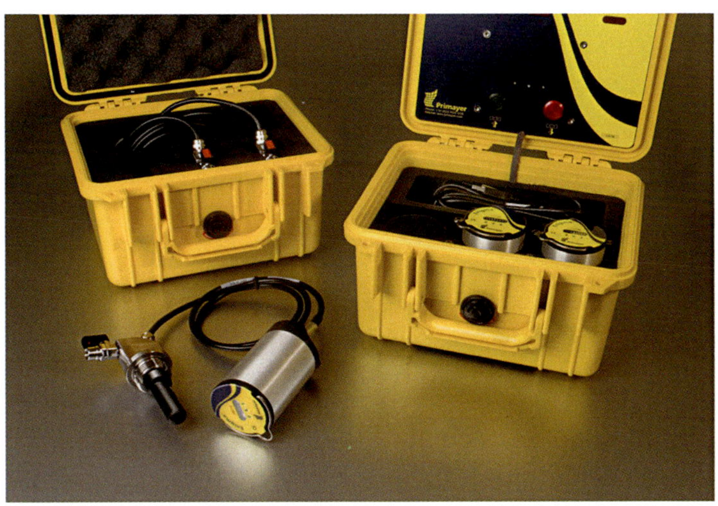

Figure 4.9 Correlating logger system with hydrophones (*Source*: Primayer).

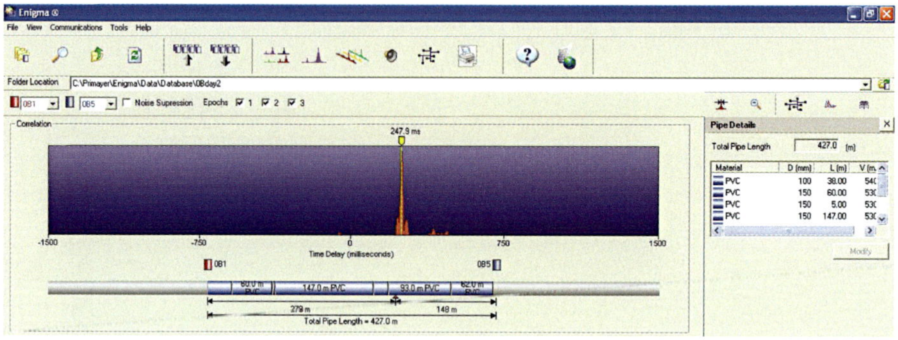

Figure 4.10 Example from map at Figure 4.7 on PVC pipe with leak located over 427 metre (*Source*: Enigma, version no. 1.4.11.0, Released 2010, Primayer, Denmead, UK).

4.5.1.4 Advantages and disadvantages of multi-point correlating loggers

Advantages

- Correlation can be pre-programmed and thus usually carried out at night when ideal acoustic conditions exist, because;
- Water pressure is higher (assuming no pressure control) and thus leak noise propagates much further
- Background acoustic noise is lower
- Multiple correlations carried out (at say one hour intervals) to clearly separate consumer use from (continuous) leak noise
- One leak may be detected by more than two loggers thus allowing accurate location, without need to measure the velocity of sound.
- As loggers are deployed at multiple points along a pipe or network, multiple leaks may be located (assuming they exist).
- Improved sensitivity when detecting low level (difficult) leak noise.
- Operating distance not limited by radio transmission range.

Disadvantages

- Slower to use than radio based correlation, especially if deployed overnight
- If a correlation result is not obtained then more time consuming to 'play' with filters on-site
- Will detect constant 'non-leak' noises

4.5.1.5 Noise loggers with correlation mode

These are hybrid devices which can function simultaneously as a noise logger and multi-point correlating logger. The noise logging feature is used to identify constant noises during the logging period and the correlation feature is used to locate the position of the leak.

4.5.1.6 Advantages and disadvantages of noise loggers with correlation mode

Advantages

- The logger can be used in drive by data collection, lift and shift mode or permanently deployed
- The noise logging function will pick up leak noise that can only be heard by one logger

- Correlation can be pre-programmed and carried out at night when ideal acoustic conditions exist
- Multiple correlations can be carried out to separate consumer use from (continuous) leak noise.
- As loggers are deployed at multiple points along a pipe or network, multiple leaks may be located (assuming they exist).

Disadvantages

- Unlike correlating loggers (described in 3.2 above) they do not use "correlation" to detect leaks. They are thus less sensitive when operating on difficult leaks.
- On some models to switch them into location mode requires a site visit to each logger (thus potentially easier to take portable correlation device to site).
- Correlation processing is only carried out for approximately 15 seconds which can be insufficient to correlate "difficult" leaks (portable correlation devices and multi-point correlating loggers execute correlation processing for longer and thus have increased performance capability).
- Slower to use than radio based correlation, especially if deployed overnight.
- If a correlation result is not obtained then it is more time consuming to "play" with filters on-site.
- Will detect constant "non-leak" noises.

4.5.2 Sources of error in correlation

4.5.2.1 Knowledge of pipe network

A thorough knowledge of the layout of the pipe network is required to ensure that the correlator is making measurements on the correct section of pipe. Where any doubt exists then the pipe must be traced using pipe location equipment. If there is any error in the location of the pipe, resulting in the incorrect distance between sensors being entered into the correlator, then this will result in an erroneous leak position (dry hole).

If the leak is located outside the length of pipe between the sensors, then the correlator will ignore the transit time from the leak to the nearest sensor, as this transit time will be the same for both sensors. This will give the result that the leak will appear to be located directly at the sensor nearest to the leak. This is often termed "out-of-bracket" (See Figure 4.11). In this situation a sensor must be moved towards the leak to continue the leak location exercise.

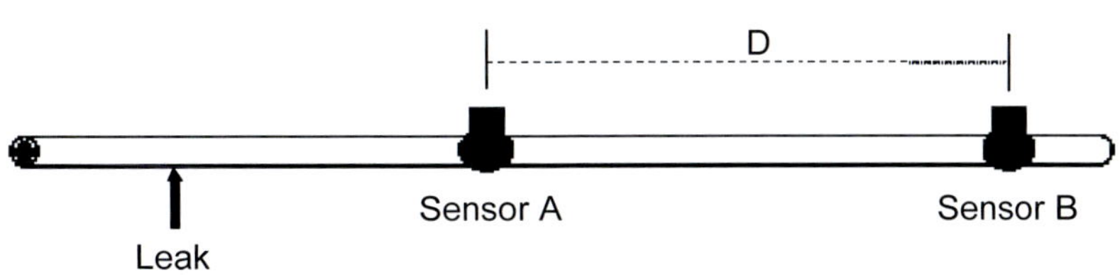

Figure 4.11 Out-of-Bracket Leak (*Source*: Primayer).

If there is a on a pipe that connects to the main pipe, then the leak noise will appear to spread from the point of connection and so will appear as a leak at that point. In this situation it is necessary to move one sensor to the connecting pipe (and so place the leak between the sensors) in order to locate the leak accurately. Figure 4.12 shows a schematic for this situation.

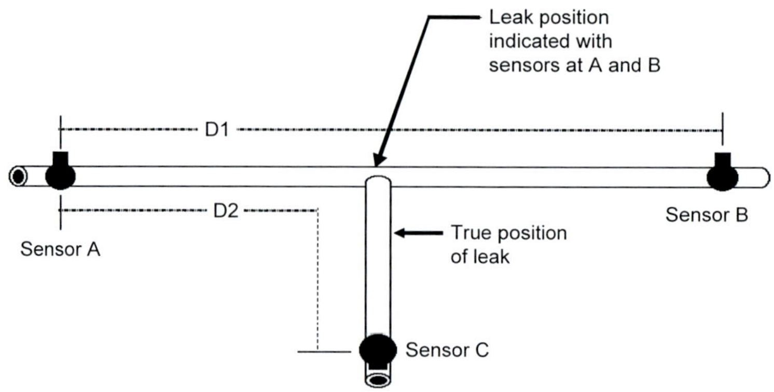

Figure 4.12 Leak on Connecting Pipe (*Source*: Primayer).

4.5.2.2 The sound velocity problem

The accuracy of leak noise correlation is highly dependent upon the value of sound velocity used. If the leak is equidistant from both sensors then the leak position result is unaffected by the velocity. The error in leak position, due to use of incorrect velocity, increases as the leak position becomes closer to either sensor and this is often the largest single reason for dry holes. On plastic pipes and on large diameter mains the sound velocity varies much more from the theoretical values (as compared to the variation on smaller diameter ferrous mains). Use of an accurate velocity value is also important when working over longer distances on all pipe materials.

All correlators will store theoretical values and these are the values used on a first correlation run. It is very strongly recommended that:

(a) The velocity of sound is measured by introducing an artificial noise (such as opening a fire hydrant). This noise can either be between the sensors or, an out-of-bracket noise. There may be difficulties in finding a suitable noise source or in detecting it given that the leak noise is also present.
(b) Carry out two correlation runs (on the leak). This is illustrated below (in fact it simply gives a more accurate leak position and the value of velocity is given as a by-product). Two correlation runs are carried out from three sensors. These should be spaced reasonably well apart and the distance between sensors must be correctly entered before commencement. This can easily be done between three correlation loggers if they are already deployed.
(c) Linear regression method. This gives the optimum result and is an extended form of the method at (b) above. One sensor remains static while the second sensor is moved to undertake a minimum of three correlations at different distances. Each pair of time delay and distance measurements is saved. Time delay against distance is graphed giving a result with high confidence (Figures 4.13–4.15).

4.5.2.3 Location of non-leak noises

Correlation will give a location for any noise generating frequencies within the bandwidth of the correlation system. Non-constant noise, such as consumer water use, can be a cause of misleading results. When using a radio-based correlator the operator must have sufficient skill to identify consumer use and repeat the correlation exercise some time later to eliminate this possibility. Correlating loggers separate non-constant noises from constant noises by performing a number of correlations separated by typically one hour to ensure the noise source is always present (a characteristic of leak noise). Correlation will

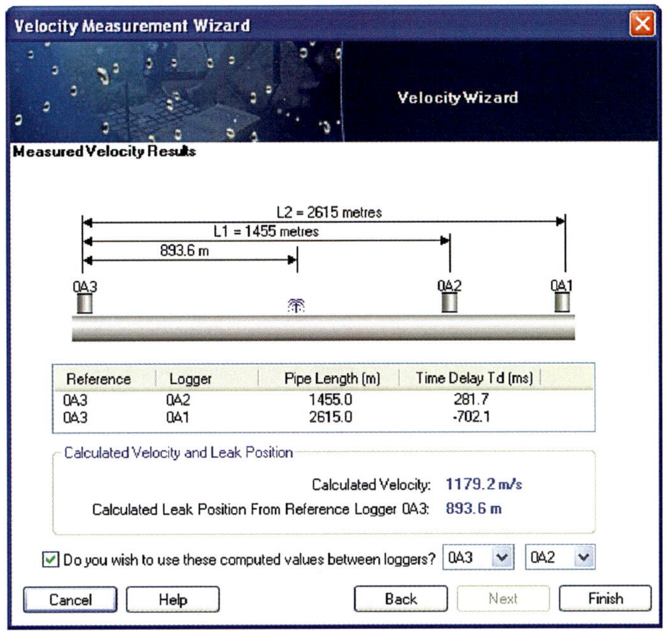

Figure 4.13 Velocity measurement with two correlation runs (*Source*: Enigma, version no. 1.4.11.0, Released 2010, Primayer Ltd, Denmead, UK).

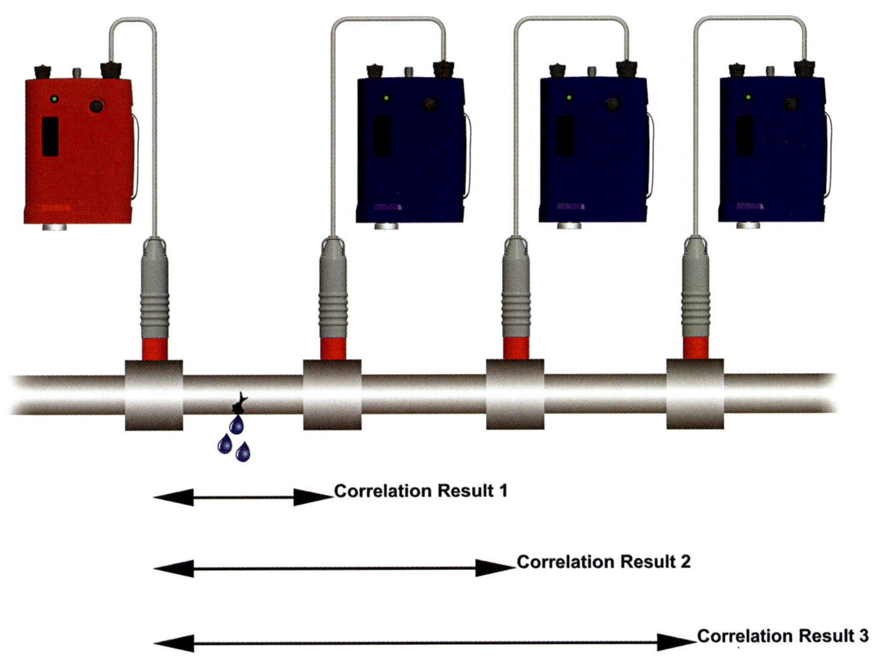

Figure 4.14 In this example the blue sensor is moved to perform three correlations (*Source*: Halma Water Management).

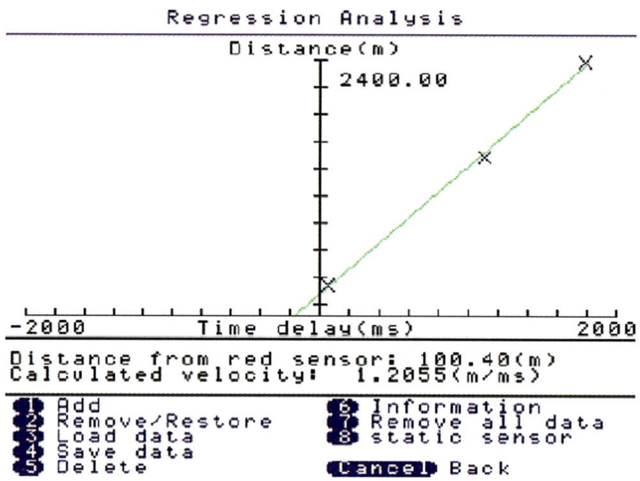

Figure 4.15 Time delay v distance is plotted to give accurate leak position and velocity of sound (*Source*: Halma Water Management).

also give the location of constant non-leak noises such as noises generated by pumps, mechanical meters, PRV's turbulence at bends, and so on. Again training of operators is important to be able to identify these as non-leak noise sources.

4.6 METHOD E: IN-LINE LEAK DETECTION TECHNIQUES

Specifically designed for large diameter pipes, in-line systems are able to discriminate between multiple leaks in a single length of pipeline. Pipelines can be inspected while under pressure and in service, and leaks are accurately located. Equipment is suitable for potable water applications, and no disruption to customers is needed.

There are two types of in-line systems; tethered and free swimming. Both have their advantages and disadvantages, but in both cases a sensor passes directly beside leaks, meaning that neither the pipe material nor the type of leak is relevant. The proximity of the sensor to any leak also results in very sensitive systems and the detection of small leaks.

4.6.1 Tethered systems

Tethered systems operate by deploying a hydrophone into the pipeline to be inspected. The hydrophone is connected to a signal processing and display unit via an umbilical cable. The sensor travels along inside the pipe pulled by the flow of water acting on a drogue (parachute) attached to the front of the sensor. As the sensor passes any leak on the pipeline it will detect the noise being generated by the water escaping through the leak. At this point the operator is able to stop deployment of the sensor (by stopping deployment of the umbilical) and then position the sensor at the leak position by withdrawing or deploying the umbilical as necessary.

Once the sensor is sited at the leak, the position of the sensor can be determined using a locating system mounted in the sensor head. A second operator can track the position of the sensor head during deployment using this locating device giving an accurate indication of the sensor location and pipe track. Having pin-pointed the position of the sensor, the exact location of any leak can be marked on the ground over the pipe (Figure 4.16).

Figure 4.16 Tethered system access through 48 mm or above connection and only goes with the flow (*Source*: Pure Technologies and JD7).

Tethered systems are best suited to work in relatively straight pipelines where deployments of up to 2 km from a single insertion point are possible; other constraints may restrict the length that can be surveyed from each insertion point. Careful planning of the work will maximise the distance that can be surveyed.

Tethered systems are deployed through a tapping made into the pipe. Typically these insertion points may be air-valve connections, insertion flowmeter tappings or specially installed points.

Tethered systems that can access the main through fire hydrants are also available. Alternative sensors, including video and ultrasonic pipe-wall inspection, are also available (Figure 4.17).

Figure 4.17 Push tethered system which can be operated with or against the water flow with access through tapping point or fire hydrant (*Source*: JD7).

4.6.2 Free swimming systems

Free-swimming data acquisition systems for detecting leaks in fluid pipelines are also available. These systems are inserted into a live pipeline and pushed along the pipeline by the water flow. At the end of the inspection a net or similar capture device is used to catch and extract the system from the pipeline or they are discharged into a open catchment area such as a reservoir and recovered from there an acoustic recording is made during the entire inspection.

Free swimming devices are also available that during the free flowing inspection the system has the capability of capturing CCTV footage and acoustic noise during the same survey.

Leaks are identified during data analysis following the removal of the system. These systems consist of the following components:

(a) Some of the free-swimming sensors contain all or just some of the following :- an internal CCTV camera, acoustic sensor, tracking sensors, acoustic transponder, data processor, memory device and batteries.
(b) Above-ground tracking devices (which are used to track the progress of the sensor through the pipe).
(c) Insertion equipment/launch tube.
(d) Retrieval equipment.

The maximum length of pipeline that can be surveyed is determined by the flow rate in the line. For instance, with a flow rate of 1 m/s and a maximum operating life of 12 hours, the system can survey 43 km from a single insertion point, some of the available technology can record for 24 hrs so hence this distance can be doubled however it would be very unwise to complete a survey of such distance. The system can traverse around tight bends and through inline valves (Figure 4.18).

Figure 4.18 Free-swimming Sensor Deployment Schematic (*Source*: Pure Technologies).

The on-board instrumentation allows the velocity of the system at all points along the survey route to be calculated during post-processing. This, combined with the use of above ground tracking devices, allow for the accurate location of any leaks (Figure 4.19).

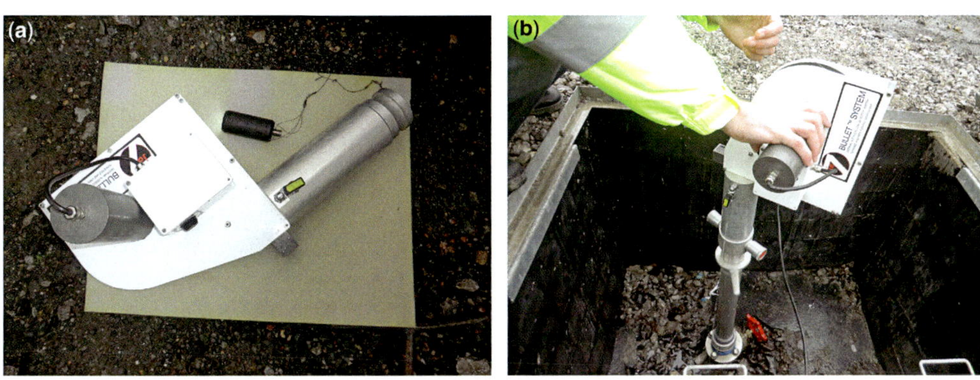

Figure 4.19 (**a**) Free flowing device combined CCTV, acoustic head, tracking device and launch tube (**b**) Free flowing device launched into main (*Source*: JD7 Ltd).

4.7 METHOD F: NOISE LOGGERS – NON CORRELATING

Previous sections show how noise is created by a leak and propagates through a pipe. Leak noise loggers are designed to "pick up" this leak noise by being placed on available fittings, usually with a magnetic coupling (Figure 4.20).

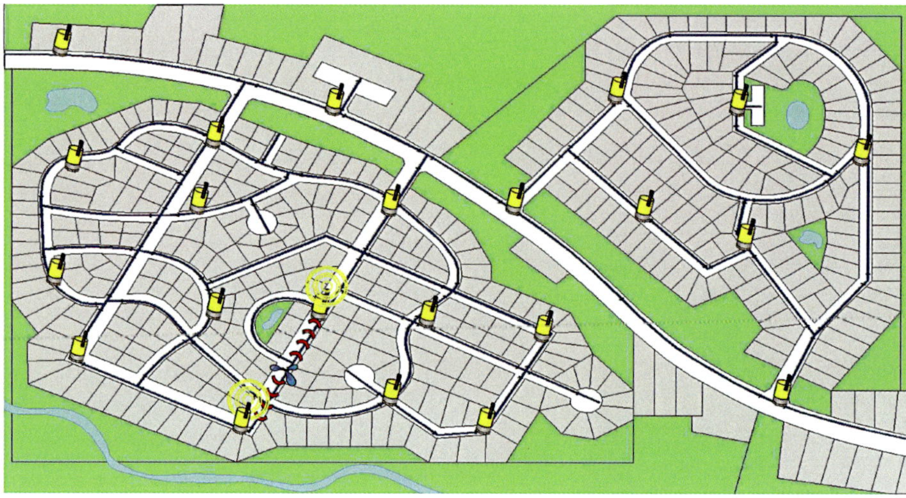

Figure 4.20 Noise logger schematic (*Source*: Halma Water Management).

The leak is identified by each logger unit individually based on the noise signature of a leak being consistent and loud against the background noise. Typically, measures of noise and consistency together with a graphical representation are supplied to the operator. The leak position is "localised" to being between two loggers for follow up pinpointing. The objective is to survey large areas at low cost to maximise efficiency in an active leak detection strategy (Figure 4.21).

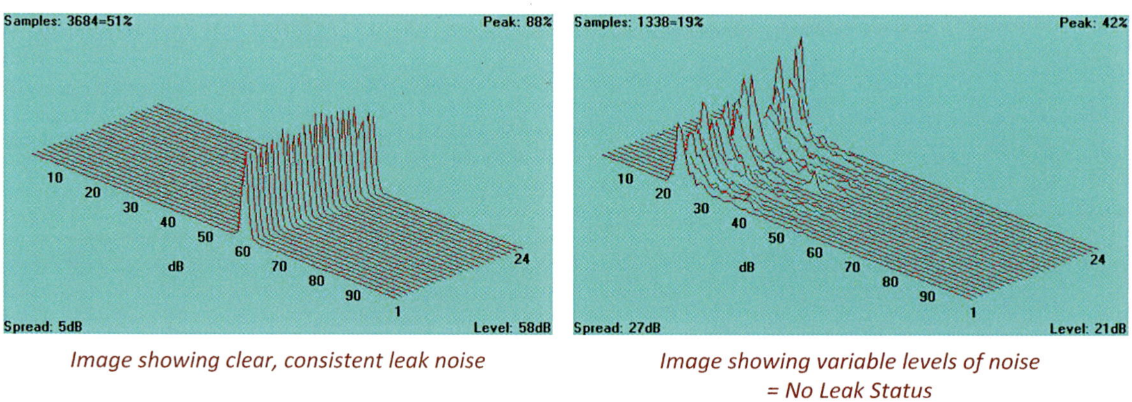

Figure 4.21 Typical leak/no leak display (*Source*: Halma Water Management).

Loggers are usually programmed to log during the middle of the night where interfering noise (traffic, legitimate water usage, etc) is at a minimum and leaks can be most easily "heard". Exact sampling regimes vary across the available systems.

Early systems were manually programmed and downloaded with the operator determining whether or not a leak was present from data supplied. As mass deployment evolved the need to further automate and ease the process became apparent with the objective being to survey large areas quickly and "automatically" at low cost.

To the above end units with radio download and automatic leak determining algorithms were introduced in the early 2000's. Many such systems are now available. Unit cost has reduced drastically with volume and technological evolution and units are now deployed in large numbers with rapid cost effective surveys possible.

Multiple deployment methodologies have evolved to suit operating requirements as follows:

- Direct download
- Drive-by patrol
- Lift and shift
- Permanent installation

Each of these is described in the following sections.

4.7.1 Direct download

Usually where smaller numbers of loggers have been installed and requiring manual download. This is commonly used where a specific problem is being investigated and the loggers will be removed after the problem has been located.

4.7.2 Drive by patrol

Loggers with a radio download facility are placed on site and their download device vehicle mounted for rapid drive by sweeps of an area. This can be used in two ways.

4.7.2.1 Fixed

Where an area has the need for frequent surveys the loggers are left in place for immediate survey whenever required. In some instances thousands of units are deployed in this mode, particularly where leakage is being targeted in "open" networks where flow base zones (DMA's) are not in place.

4.7.2.2 Survey

Loggers are deployed in an area (often a problematic DMA) so that it can be surveyed and brought under control. Interestingly, experience shows that several sweeps are required to optimise leakage as initial sweeps reveal leaks that are masking larger, quieter leaks that are identified by subsequent downloads after initial leaks are repaired. When all leaks have been identified the loggers are moved to a new target area (Figure 4.22).

4.7.3 Lift and shift

In order to survey very large areas quickly the industry has developed a "lift and shift" methodology where large numbers of loggers are moved daily. Data is downloaded automatically via radio during the process. Hand held retrieval units directly or remotely transfer the leak list to a central office for pinpointing follow up.

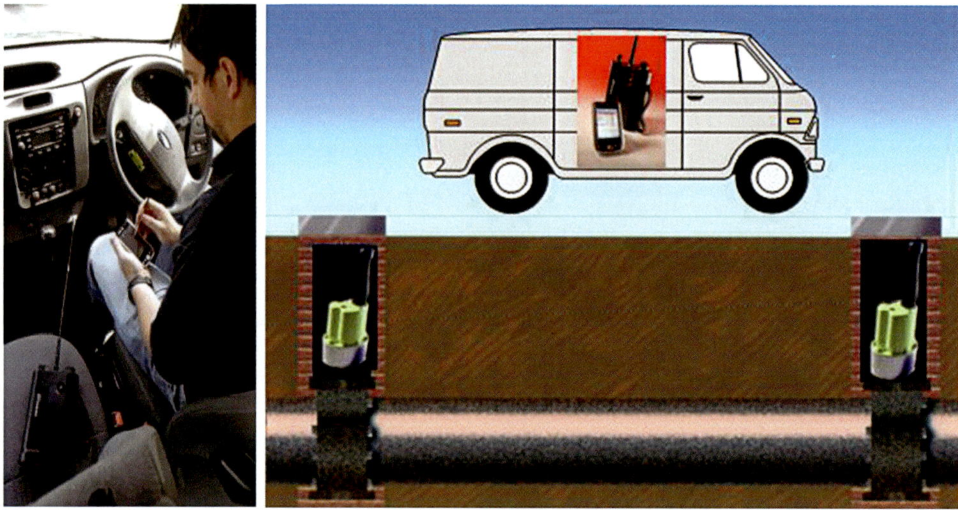

Figure 4.22 Drive by Patrol (*Source*: Halma Water Management).

Technology has evolved to suit the methodology with GPS positioning, mapping and other aids enabling a highly efficient operation to be carried out (Figure 4.23).

Figure 4.23 Lift and Shift Deployment and GPS Mapping Function (*Source*: Halma Water Management).

4.7.4 Permanent installation

Modern communication technology (SMS, GPRS, Radio) and reducing costs of data transfer now means that noise logging can be economically installed as a field network. Various data transfer methodologies are available and common download platforms with AMR systems have also been introduced.

The following areas are particularly suitable for permanent installation.

- Areas with a high burst frequency
- Areas that are traditionally difficult to survey (i.e. town centres and main roads)
- Areas with no DMA structure in operation where acoustic noise logging offers a cost effective alternative.
- Previous DMA "hotspots" where ongoing survey is required to ensure leakage levels remain manageable.

Leak noise data is automatically transferred to the central monitoring station. For ease of running data is often linked with GIS or mapping systems, to provide a quick pictorial overview of the network (Figure 4.24).

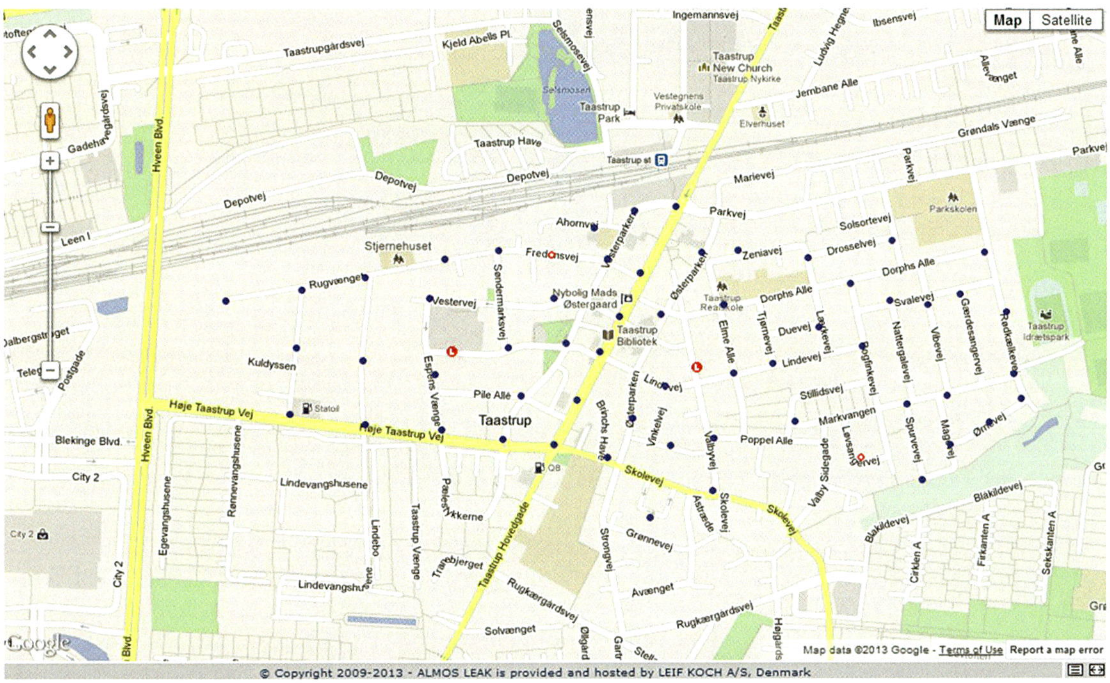

Figure 4.24 Permanent noise logger installation by street map only or full aerial view (*Source*: Halma Water Management).

The availability of immediate alarms reduces leak run time to a minimum and, with effective follow up repair, provides a huge dividend in water saved.

Combining leak alarms with DMA flow data enables effective leak sizing to prioritise and optimise follow up activity. Survey labour expense is removed with leak detection becoming a more specialised pinpointing and repair activity.

The benefits of the fixed network approach are evident. With technology providing ease of installation the decision to deploy (or not) is now largely down to an economic comparison of labour costs against capital investment, and the additional savings and benefits of immediate notification.

4.7.5 Conclusion

Leak noise logging has evolved in scale and technology to provide multiple methodologies for leak localisation. Consideration of labour costs, rate of intervention required, status of the network and availability of DMA's (flow zones) will allow strategy of deployment to be optimised (Figure 4.25).

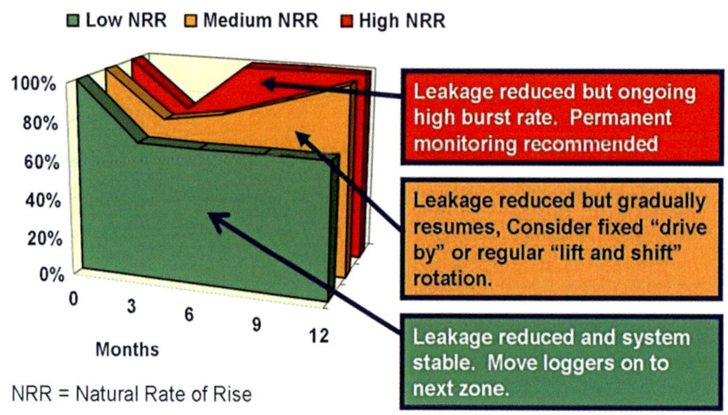

Figure 4.25 Theoretical Model for Leak Noise Logging Methodology Selection (*Source*: Halma Water Management).

4.8 METHOD G: ELECTRONIC AMPLIFIED LISTENING DEVICES

When a pressurised water pipe develops a leak the water flows out into the surrounding ground at high speed, which causes the pipe and soil to vibrate at the exit point. This sound, or vibration, is transmitted by the pipe (structure borne), the surrounding material (ground borne) and through the water itself over a range of frequencies. Careful application of leak detection techniques will enable the operator to eliminate detected noises generated by poor pipeline design or consumer usage and to identify leakage due to pipe system damage.

In addition to being transmitted along the pipeline (both through the water and the pipe wall) the leak noise is transmitted into the ground around the pipe. The noise travels much better through "hard" materials so that the noise travels much further along metallic pipes than asbestos cement pipes which themselves are better than plastic pipes. Ground material generally provides a poorer travel path than the pipeline itself. However, usually some noise transmitted the short distance to the surface. Soft sandy ground provides a worse travel path than well compacted ground with a hard paved surface covering (Figure 4.26).

Factors producing good quality leak noise include high water pressure, hard backfill, a small rupture, clean pipes, metallic pipes and small diameter pipes. By contrast, factors producing poor quality leak noise include low water pressure, soft backfill, split mains, encrusted pipes, soft/lined pipes and large diameter pipes.

Since "leak detection" began, operators have been "listening" for this leak noise using mechanical devices. Traditional listening sticks for detecting water leaks rely on only one of the user's senses – hearing – the experience and skill of the operator is paramount and, at best, users are only ever able to detect leaks that produce loud noises.

However, it must be noted that not all leaks produce a noise audible to the human ear. Contrary to common perception it is not always the largest leaks which are the loudest; often a large split in a water pipe will produce a less clear noise than a small hole. This can be particularly true in P.V.C., P.E. and M.D.P.E. pipe materials. For this reason amplifying the noise with an electro acoustic microphone is becoming increasingly important to find leaks particularly in networks where these materials are increasingly used.

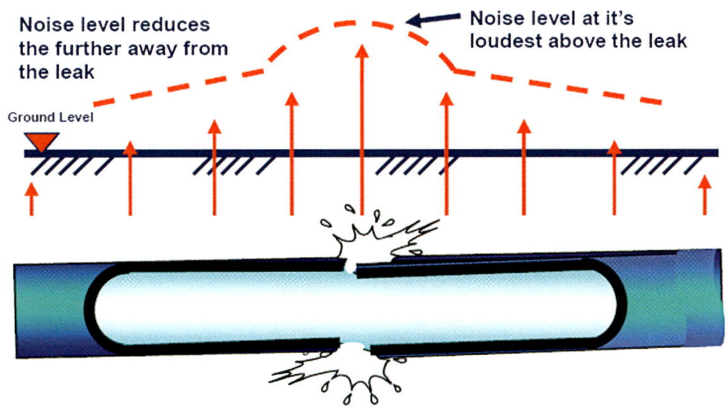

Figure 4.26 Schematic showing propagation of leak noise through ground (*Source*: Halma Water Management).

Modern electronics therefore provide the benefits of advanced sensor technology amplification and filtering to undertake this operation more effectively.

4.8.1 Operational practice

The typical key components of a modern electronic amplified listening device are shown in (Figure 4.27) below:

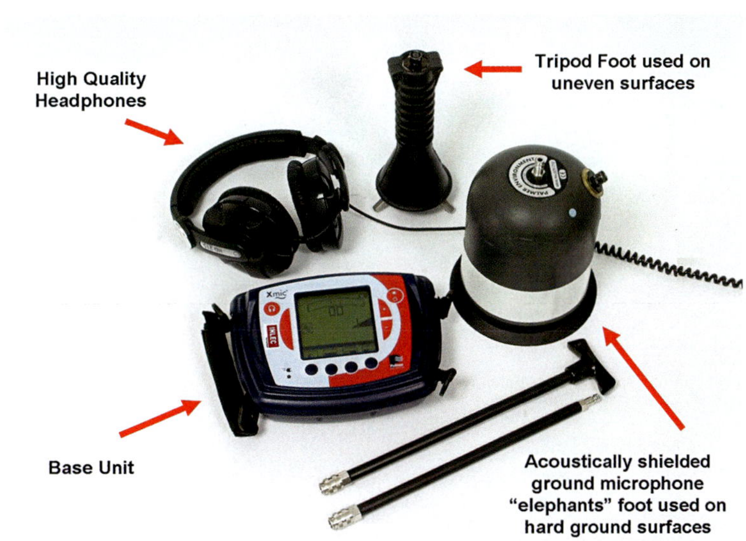

Figure 4.27 Components of Electronic Amplified Listening Device (*Source*: Halma Water Management).

Pinpointing a leak position using an electronic amplified listening device involves a process of comparing a number of leak noises. To begin with, the operator must select the most suitable sensor device; the microphone foot for hard ground surfaces or the hand probe for soft ground.

To operate the ground microphone safely and effectively, the operator must adjust the headphone volume control to a comfortable listening level. Once the noise has been heard, the headphones should be muted before moving the microphone foot or hand probe to the next test position.

The operator should repeat the sequence to listen to each of the test locations as he moves along the pipe route in the direction where signal strength is increasing. If the leak noise level falls he has passed the leak and should go back and reduce the distance between each measurement. The loudest leak noise will then indicate the location of the leak bearing in mind the ground conditions already mentioned.

These basic principles can be used in a number of operational modes.

4.8.1.1 Survey by listening at fittings – electronic listening "stick" accessory

All listening devices give better response when in direct contact with the pipe, particularly on metal pipelines.

In this mode the unit is used for locating leaks by fixing it onto a contact point or fitting, such as a valve, meter, hydrant or stop tap. A hand-probe/extension rod can be used to listen at these points and these provide good sound pick-up particularly if the pipe is metallic.

This procedure was once commonly referred to as "bashing" when used with mechanical listening rods and is used to narrow down the location of the leak, a procedure known as localisation since the leak will be "localised" to some point between the fittings. It will then be located to an approximate position and then "pinpointed" for the exact position prior to excavation.

It is important to note that when listening on pipe-fittings the location of the point of maximum noise will probably not indicate the leak position, only the fitting closest to the leak (Figure 4.28).

4.8.1.2 Survey/pinpointing by surface sounding "elephants foot" – hard ground

The ground microphone is used on the surface in the following circumstances:

- To pinpoint the leaks position after the "Localization" and "Locate" survey by listening at fittings.
- When no accessible contact points are available.
- When the pipe is of non-metallic material and no leak noise is being transmitted to available fittings.

The ground microphone is moved along the surface in regular positions following the path of the pipeline below ground with the operator noting the changes in sound amplification until the area of maximum noise level is identified.

To locate the position of an underground leak an acoustically shielded ground microphone foot (commonly known as an "elephant's foot") can be placed on the ground above the line of the suspect pipe and the readings observed. This is particularly useful on hard ground conditions and provides isolation from airborne noise interference which allows it to be used in noisy and windy conditions. On soft ground it is usual for a spike or tripod sensor to be used. A hand probe with a magnetic contact can also be used to provide excellent acoustic coupling to fittings, helping to ensure clearer, louder leak noise.

Figure 4.28 Leak position from sounding fittings (*Source*: SEBA KMT).

It is important to note that the noise level will also appear stronger where there is less thickness of ground or other material for it to pass through. The leak noise will always follow the path of least (acoustic) resistance (Figure 4.29).

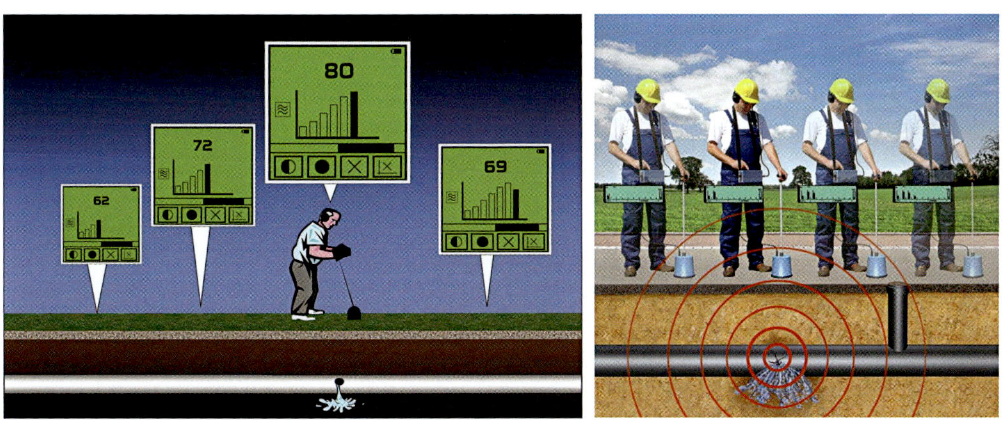

Figure 4.29 Surface sounding leak position (*Source*: Halma Water Management and SEBA KMT).

4.8.1.3 Operational efficiency – survey vs. confirmation

Whilst it is recognised that the electronic ground microphone can be used for a complete leakage survey and pinpointing operation, using ground microphones in a survey operation is very labour intensive and involves operators covering (walking) large distances. Other modern technology is now available to survey large areas more quickly and efficiently.

Specifically, these methods include:

- Measuring flow into a self contained District Metered Area and deducting known domestic and commercial usage in order to calculate the amount of real losses (leaks)
- Narrowing down the area by means of isolating parts of the distribution system (a procedure known as Step Testing); and
- Deploying acoustic noise loggers along the pipeline within a District Metered Area to identify whether a leak may be present.

The above survey procedures are far more effective than using ground microphones to identify potential leak locations.

The modern Ground Microphone is therefore primarily used today as a confirmation of correlation results before excavation. This avoids errors due to unknown pipe characteristics, or operator error leading to dry holes or unnecessarily large excavations, and should be viewed as an essential final confirmation.

The illustrations below (Figure 4.30) show how the electronic ground microphone is used effectively as a pinpointing tool prior to excavation, with the histogram of sample noises providing a clear visual depiction of the precise leak location.

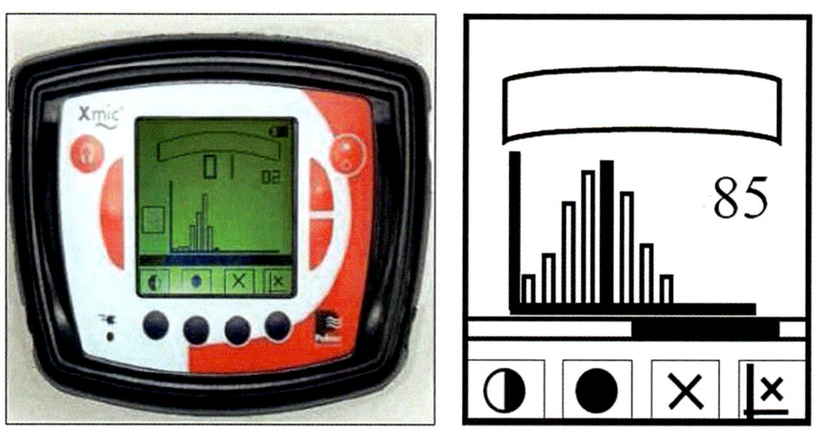

Figure 4.30 Images from Electronic Ground Microphone (*Source*: Halma Water Management).

4.8.1.4 Use where poor noise transmission along the pipe renders other techniques ineffective

In certain situations (non metallic pipes, poor pressure, holes in pipelines) leak noise may not travel over any distance along the pipelines and therefore pipe contact techniques may not be effective. In these circumstances the noise may still be sufficient at the surface above the leak to be identified by a Ground Microphone survey, and this may be the only practically effective acoustic technique.

4.8.2 Advanced features

4.8.2.1 Filters

Filters are required to selectively target and amplify leak noise and to suppress unwanted background noise and interference.

"Targeting" leak noise – Metal pipes generally produce higher frequency leak noise as do small fractures, high pressures and hard compacted ground. Modern filtering allows the operator to set the instrument to focus on the leak noise and amplify it "out of" the background interference. Filter selection can be a great help, but it must be noted that unwanted noise may have a similar frequency to that of the leak, and not all ambient noise can be isolated out. For example, traffic and machinery noises often occur in the same frequency bands as leak noise and can travel for considerable distances through both air and ground material.

For this reason it is sometimes advisable to use acoustic leak detection techniques at night when interfering noises are less.

The diagram below shows examples of approximate filter settings for sample pipe materials but please note that these are only given as a guide. The frequencies received will also depend on the ground conditions and type of leak and the filters may require further adjustment to remove unwanted background noise (Figure 4.31).

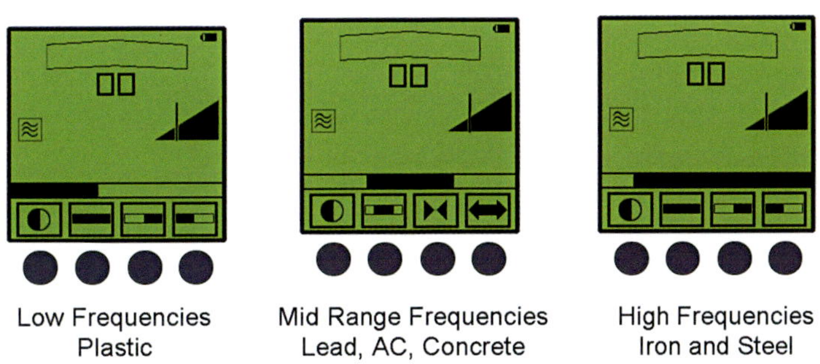

Figure 4.31 Filter settings for pipe materials (*Source*: Halma Water Management).

4.8.2.2 Memory comparison

Unlike its mechanical predecessor, the electronic amplified listening device enables automatic comparison of small noise differences to support the human ear.

Some advanced devices feature a minimum noise level memory to aid the operator to establish exactly the highest point of leak noise. This can greatly assist the operator to pinpoint the exact leak position, with comparison not possible by the human ear.

As the operator steps along the line of the suspect pipe listening to the sound levels, the device automatically records the base level of noise by checking and memorising the lowest noise level. This is the constant background (leak) noise. As the sensor is moved, a series of readings are taken. These are displayed digitally and graphically to show the difference between each reading clearly, for simple and precise pinpointing.

The illustrations below (Figure 4.32) show that the memory comparison function clearly identifies that the leak is located at the highest point of the histogram.

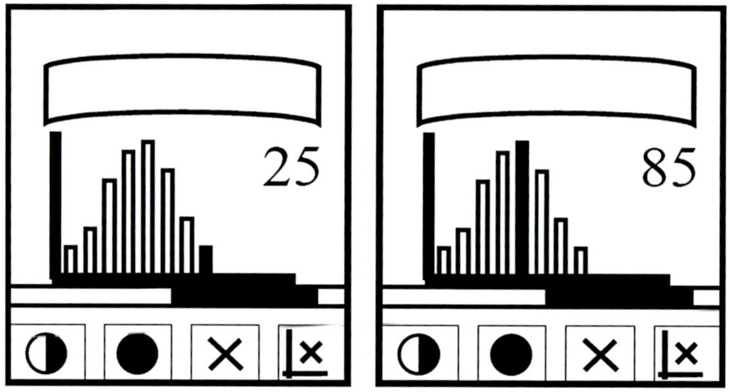

Figure 4.32 Ground Microphone display (*Source*: Halma Water Management).

In this example, eight consecutive readings have been taken by stepping the ground microphone along the length of the pipe and recording the leak noise at each position. This data is recorded and displayed as a vertical bar on a histogram. Each time a sample is taken, a relative number appears on the right of the screen. Each sample can be subsequently selected to check the relative number. This indicates the greatest leak noise acquired and the probable position of the leak.

4.8.2.3 Amplification

Using the latest acoustic technology, modern electronic listening devices amplify signals which otherwise could not be heard by human ear, making them far more effective than mechanical listening sticks. The output of the amplifier is typically fed to high quality headphones to enable the operator to better detect subtle variations in frequency or pitch that enable the leak location to be positioned accurately.

The most effective portable amplifier modules have an integrated LCD display that enables noise levels and dynamic sensitivity (signal strength) received by the microphone to be displayed graphically as well as audibly.

4.8.3 Advantages and disadvantages

We can summarise the operational advantages and disadvantages as follows:

Advantages

- Low cost
- Easy to use
- Effective in some circumstances where techniques reliant on noise transmission through the pipe are not effective.
- Provides effective leak confirmation following correlation.

Disadvantages

- Very labour intensive
- Inefficient when used as a surveying tool.
- Difficult to operate effectively in high-noise conditions and busy urban environments which may require working at night.
- Performance subject to soil conditions.

4.8.4 Conclusions

The electronic amplified listening device is an essential tool in the leak detection armoury. The most modern devices include advanced features to help the operator to survey, pinpoint and confirm leakage positions.

Whilst electronic devices with ground microphones are still used routinely in many parts of the world to survey areas of suspected leakage, its usage has evolved with the introduction of new areas of technology. Today, equipment exists that enables the leak detection operator to survey large areas much faster and utilizing significantly less manpower than that required by ground microphones alone.

However, the modern ground microphone is invaluable in general leakage work and particularly useful when used as a final confirmation of the leak position detected by a correlator prior to excavation, thereby greatly reducing the number of dry holes or unnecessarily large excavations (Figure 4.33).

Figure 4.33 Electronic Amplified Listening Devices (*Source*: Primayer).

4.9 OTHER TECHNIQUES

Several other techniques have been developed as alternatives to the acoustic methods described above.

4.9.1 Thermal imaging

Thermal infrared imagers are detector and lens combinations that give a visual representation of infrared energy emitted by all objects. In other words thermal imagers let you "see" heat. Depending on the sophistication of your system, thermography is capable of providing very detailed images of situations invisible to the naked eye.

Thermal Infrared pipeline surveys not only provide data on possible leakage points but also on the status of the wayleave (right-of-way), showing up any ground disturbance or building works over the pipeline.

The surveys generally fall into two categories:

4.9.1.1 Low level surveys

These are surveys of the known course of a pipeline over open country. By carrying out a low level survey, extremely high resolution images may be produced which are invaluable to leakage engineers for defect identification and prioritization.

4.9.1.2 Higher level surveys

These are surveys of rural countryside where pipes lie but whose exact position is not known. These are usually flown at about 600 m (2000 feet), which gives a good balance between resolution and flying time (and therefore cost). The pipe or pipe bed is usually visible around 60% of the time depending on ground composition and cover.

Although infrared equipment is a valuable diagnostic tool, it merely provides a "map" of radiant energy. It cannot for example give a definitive answer to why a particular area is at a certain temperature or radiating at a certain emissivity. Consequently skilled interpretation can be a valuable source of advice in this area.

High resolution thermal imaging has proven to be a versatile technique for identifying pipeline (or reservoir) anomalies in rural areas in addition to identifying environmental effects such as discharge into watercourses.

The pipeline is flown over in a series of tracks, the number of which depend on the pipeline route. Each of these is related by an on-screen time stamp to the real time video data. This map is then marked with any thermal anomalies noted and used as the basis for discussions with pipeline management staff that are familiar with possible valid causes for many of the anomalies (such as pipe infrastructure, fitments, valves etc. e.g.). This is often provided in parallel with a tabulated list of anomalies, their track number, time/date stamp, classification of priority (high, medium, low) and appropriate comment. The technique does require considerable man hours in analyzing and tabulating the data.

4.9.2 Ground Penetrating Radar

GPR (Ground Penetrating Radar) is so named because the radar is able to produce an image of what is below the ground by reflecting radar frequency waves emitted by the transmitter from any interface within the ground, this being earth/water, earth/rock, rock/air and so on back to the receiving antenna. Usually this antenna is part of the same unit as the transmitter which is drawn over the ground at a slow speed producing data which is processed and converted into vertical cross sectional slices of the ground.

This image is shown on a screen for the engineer to interpret the results and decide whether a leak is present. This requires an experienced engineer to decipher.

The depth of vision can be as much as 3–5 m but this is in perfect situations which are rarely seen (Figure 4.34).

There are several European Union co-funded research projects on technology currently in progress – with varying results and successes. Two such projects are "Waterpipe" and "Leaking". The projects have similar objectives – to provide a non-intrusive leak location technology:

"Waterpipe" is a system where the leak is located by Ground Penetrating Imaging Radar – GPIR. The objectives of the project were to investigate and develop a high resolution imaging ground penetrating radar for the detection of pipes, leaks and damages to underground infrastructure – and to provide imaging of the damaged region. A further enhancement was to produce an integrated system that will contain both the GPIR equipment and a Decision-Support-System (DSS) for the rehabilitation management of the underground water pipelines. This would use input from the inspections to assess, probabilistically, the time-dependent leakage and structural reliability of the pipelines and a risk-based methodology for rehabilitation decisions that considers the overall risk, financial, social and environmental criteria.

"Leaking" had objectives to investigate and develop an innovative leak inspection equipment for water pipelines based on microwave technology (a Continuous Wave Doppler radar, a Frequency Modulated Continuous Wave radar and a radiometer), and a decision support system, that stores available data on the pipe network, and receives input from leak inspections. It should be able to perform condition assessment to determine residual life time of the pipeline in question.

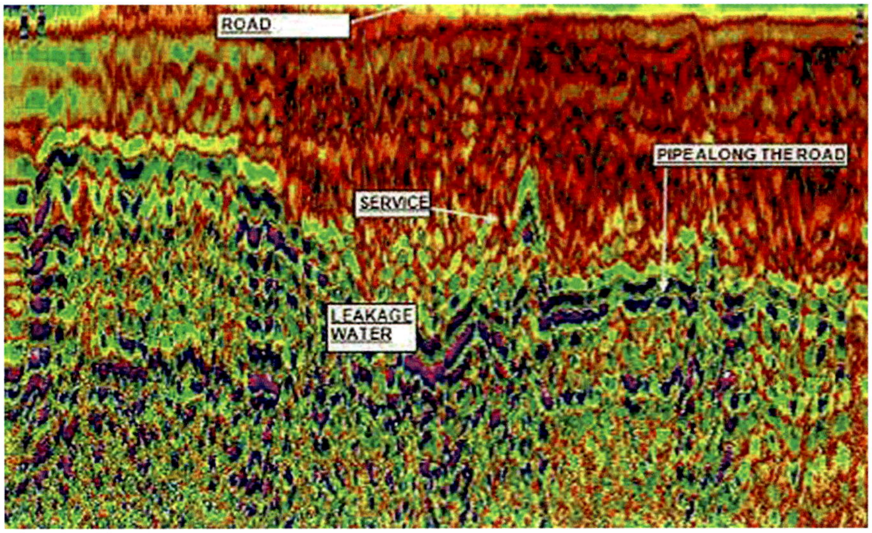

Figure 4.34 Data from GPR (*Source*: R Brier).

4.9.3 Ultrac method

The Ultrac methods are listening methods and can be seen as a further development step along the line of the listening stick, using electronic signal processing. They have initially been developed for use on plastic pipes and pipes of large diameters, although they will also work with normal steel pipes. The intention is to provide a monitoring device that works when other methods fail, as for example can be the case with correlators on plastic pipes where the sound attenuation is high. In all cases, one single hydrophone is used as transducer, normally mounted on a hydrant.

The first method uses a frequency shift upwards (pitch shift) in order to make the existing very low frequencies audible. The shift is typically from around 30 Hz to around 250 Hz. In this way, it is possible to judge by hearing from the typical noise pattern whether a leak exists or if there is only a background noise. The distance to the leak can be determined by comparing the high frequency components of the noise to the low frequency ones, since the high frequency components suffer more attenuation with distance than do the low frequency ones.

The second method can be denoted as an active one, in contrast to the first which is passive. The active feature implies sending a short acoustic pulse, and waiting for the echo from the leak. When the speed of sound is known, the arrival time of the echo is converted to the distance. This method requires an advanced internal signal processing and gives a satisfactory assessment of the distance. The working principle is similar to that used in sonar techniques to find objects in the sea.

4.9.4 Optimization tools for leak location

The water industry has accepted a hydraulic model as routine computer simulation tool to analyze the hydraulic characteristics of the system elements (pipes, pumps, valves and storage facilities). Optimization technology has been developed to enable water companies to make use of the well-established computer model for low cost localization of leaks. Although it is impossible to exactly locate the water losses or leakages in a distribution system by just using an optimization-simulation approach, case studies have shown that the method is effective in helping engineers to narrow down the possible water loss (including leakages, un-metered and illegal consumptions etc) and thus enable more efficient leakage reduction programs.

4.9.5 Optimization principle

The method requires a hydraulic model together with pressure and flow logging data from the network, which are used by an optimisation technique to identify the likely leakage hotspots that are emulated as emitters where leaks occur and depend on the local pressure. Genetic algorithm (GA) techniques have been used to search for the leakage hotspots such that the difference between the simulated and field observed pressures/flows is minimized. Engineers need to repetitively run the optimisation tool with the MNF-hour data for the same network model. Multiple runs provide an indication of most likely leakage hotspots in the network area. The essential steps for applying the optimization modelling method are as follows.

4.9.6 System evaluation

Prior to embarking on leakage detection optimization modelling, it is essential to evaluate the system behaviour by using the hydraulic model. Figure 4.35a illustrates system layout and connectivity. The DMA's supply boundary is simulated by using a reservoir with variable hydraulic heads. The inflow into the DMA is recorded as time series over 24 hours and 6 pressure loggers are used for collecting pressures from the field. An extended hydraulic simulation is performed to evaluate the existing behaviour of the system. Figure 4.35b shows the comparison between the observed inflows and the simulated flows. The big flow difference at each time step represents the water loss within the DMA.

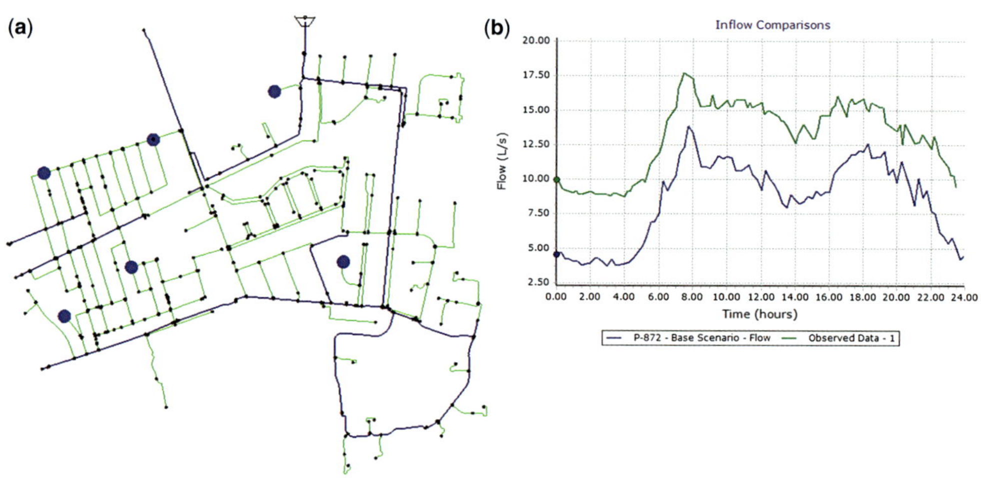

Figure 4.35 System evaluations prior to leakage detection optimization (*Source*: Bentley WaterGEMS, Darwin Calibrator Module, Bentley Systems Incorporated, Exton, PA, USA).

4.9.7 Field data process

A large set of data is collected from the field, including the pressures and the flows. The data is prepared and classified into different datasets. Each of them represents a snapshot of the system observation at the same time.

4.9.8 Optimization analysis

Leakage detection optimization study can proceed with setting up the optimization runs and then executing the runs. The software tools are flexible and powerful enough to permit engineers to use system-specific knowledge to make different target groups for effective leakage modelling.

4.9.9 Post-optimization analysis

It is important to process the identified leakage solutions, namely the identified emitter locations and the emitter coefficients. The post-optimization process includes creating leakage node map as shown in Figure 4.36, in which each identified leakage node is represented by the colour-coded node for intuitive presentation of the most likely leak areas. The colour map can be used as a good guide for field detection crew. The optimized leakage solution can be further evaluated by comparing the observed and simulated inflows over 24 hours. Figure 4.37 illustrates that the identified leakage nodes significantly improve the flow comparison.

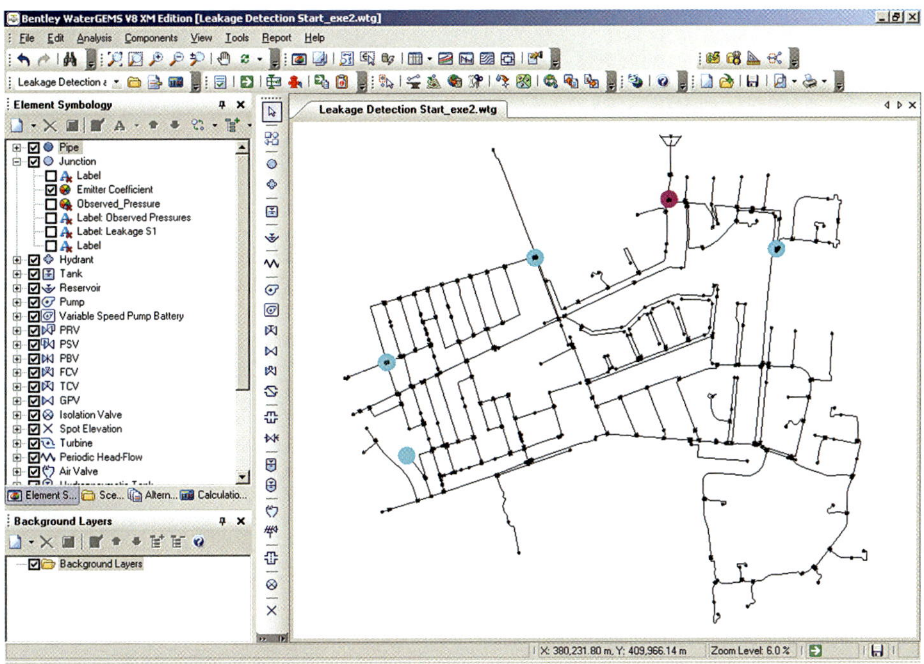

Figure 4.36 Create leakage map of identified leakage hotspots (*Source*: Bentley WaterGEMS, Darwin Calibrator Module, Bentley Systems Incorporated, Exton, PA, USA).

There are a number of alternative techniques in use. All use one or more time steps in the network model. The differences in the techniques relate to the details of the optimization methods, pressure-demand relationships, choice of time steps, the amount of demand re-allocated and the rules used for the demand re-allocation.

The location of the re-allocated demand gives an indication of leak locations or additional hidden demand. The approach will not pinpoint the leak location but will localise it, to make detection by other methods more cost-effective.

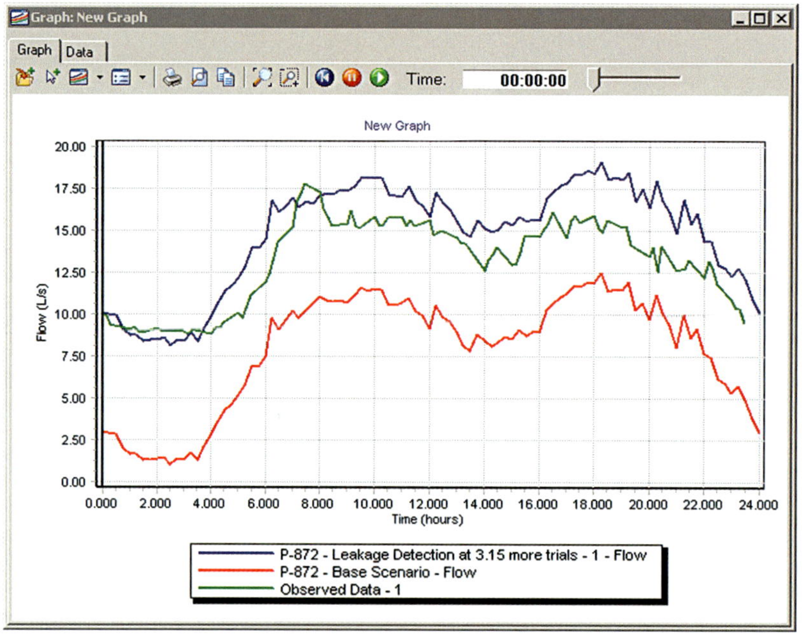

Figure 4.37 Comparison of the field recorded flows with the extended period simulation (*Source*: Bentley WaterGEMS, Darwin Calibrator Module, Bentley Systems Incorporated, Exton, PA, USA).

4.9.10 Step testing

Step testing is an effective, flow-based method of localising water loss within a zoned distribution system. It is particularly suited to identifying areas of high leakage and to use on plastic pipe materials, where leak noise is absorbed and conventional acoustic methods are less effective.

To perform a step test the inflow into a zone must be monitored. This can be achieved by deploying a data logger upon the inlet water meter to automatically transmit flow data to the operative in the network. Alternatively an additional operator can be left upon the inlet meter to manually record flow and network activity.

Once a method of monitoring has been established, then valves are closed to cut off sections of the zone known as "Steps". This demonstrates how much water is consumed in each step. Each step has an estimated customer consumption which is compared to the drop in flow at the inlet meter. If the difference between the actual drop in demand and estimated consumption is significant, this provides an indication to the operative that leakage is contained within that Step.

There are several different variations on the approach depending on the technology available, whether it is important to maintain supply and the configuration of the network.

4.9.11 Principles of step testing

- Each zone will require a plan that identifies which pipe lengths are to be used, valves to shut and in which order valves are to be shut. This should be used for all subsequent step tests, providing the zone does not change. By keeping the plan consistent the operators can provide further judgement based upon experience.

- When designing a step test plan it is important to have an optimum amount of "Steps". This will largely depend upon the size of the zone. Too few Steps may not achieve the desired reduction in leakage detection time and costs. Too many "Steps" can also be time consuming and the rate of leakage may be too insignificant for the flow meter to register.
- Another consideration when designing "Steps" is to calculate an estimation of customer consumption so that the operative has an expectation of a typical flow rate into a "Step".
- Step tests should be carried out when demand is at its lowest. This tends to be at night time between the hours of 01:00–04:00. This helps contributes to a more accurate step test as fluctuations in demand are minimised.
- Before a step test is implemented all valves required must be located on site. Once located then the integrity of the valve must be tested. This will include ensuring the valve is accessible and operable. A Zero Pressure Test (ZPT) can conclude if the valve can be closed completely without passing any water, this helps contributes to further reassurance of an accurate step test.
- There are two options in providing essential flow data to the operator closing valves in the network:
  (1) Another operator upon the inlet meter with a telephone/radio communication to the operator in the network operating valves
  (2) Radio/GPRS data logger to transmit flow data to a suitable receiving device for the operator in the network to see live flow data (Figure 4.38).

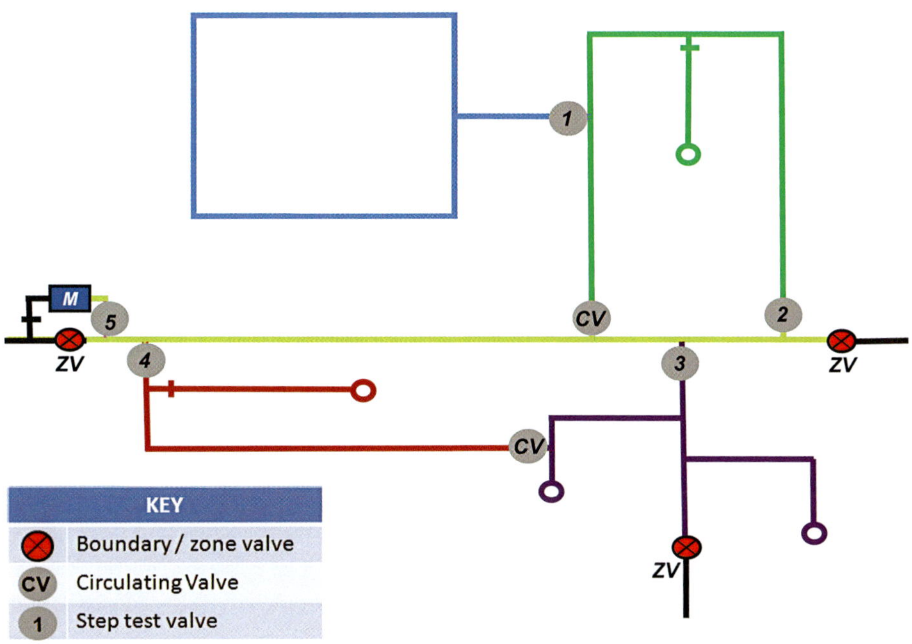

Figure 4.38 Showing a step test plan with valves labeled accordingly (*Source*: Primayer).

- There are three types of valves when operating a step test :
  (1) Valves that are permanently shut to create a zone. These can sometimes be called boundary valves or zone valves.

(2) Valves that need to be shut before the start of the test in order to create "Steps" that can be closed off during the test with a single valve, as sometimes it is not possible to shut off a section by using only one valve closure. These valves are only shut for the duration of the test and re-opened once the test is completed. They are sometimes known as circulating valves.

(3) The final type of valve is one that isolates a step from the zone. They are numbered according to the order that they need to be shut in during the test. Step 1 is typically the "Step" that is the furthest away from the meter and the last step closure is the one nearest the meter.

Figure 4.39 Graph showing results of a step test (*Source*: MAST PC Software V5.03, Halma Water Management, Cwmbran, Wales).

Step Test Valving Schedule

DMA Name	Test Area			DMA Number	555
Test Details:					
Valve Number	Operation	Time	Flow(l/s)	Comments	
Circ 1	Shut	02:00	10		
Circ 2	Shut	02:15	10		
Test 1	Shut	02:20	8		
Test 1	Open	02:25	10		
Test 2	Shut	02:37	9		
Test 3	Shut	02:51	5		

Figure 4.40 Table showing a traditional method of recording step test events (*Source*: Halma Water Management).

- It is important to allow a settling time (approximately 15 minutes) between each “Step” closure so that a stable and realistic flow rate can be obtained.
- It is good practice to record when all network events occurred, for example valve closures. So that thorough analysis, if required, can occur at a later date.

In Figure 4.39 drops in flows are shown as the “steps” are being closed. The burst was located in the final step where the flow reduces considerably. The turning on of the “steps” can be seen to the right hand side of the graph (Figure 4.40).

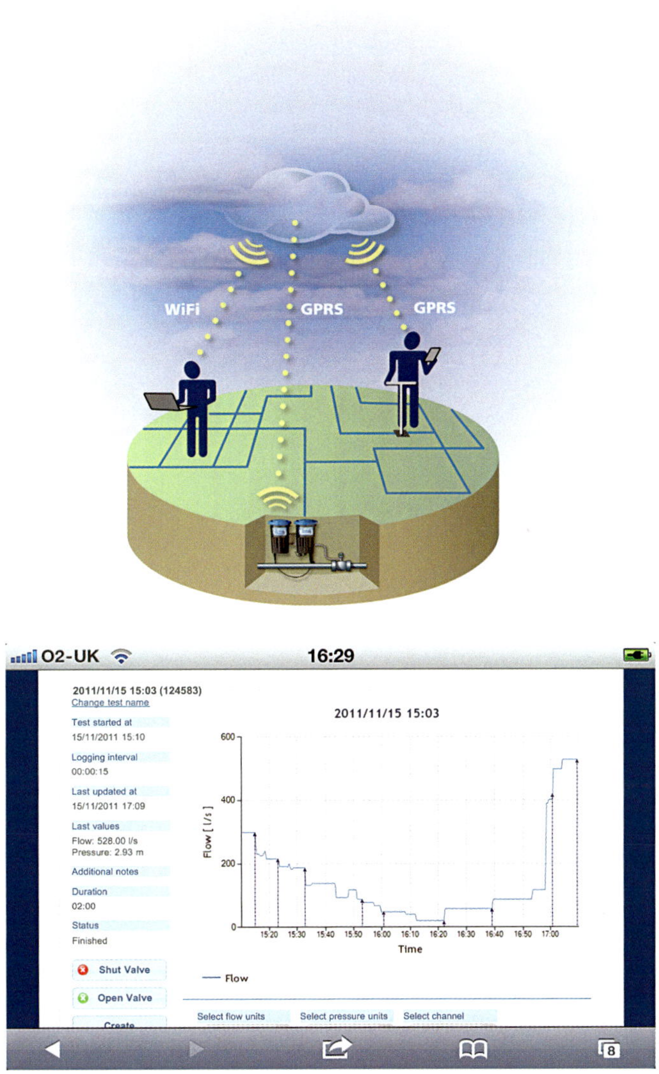

Figure 4.41 Advanced step testing (top) and data from advanced step test (bottom) (*Source*: Primayer).

4.9.12 Advances in step testing

Current technology and consumer pressures on levels of service have led to an alternative method of step-testing that avoids using predetermined steps. By obtaining real time data operators can instantly view the flow changes in the network. This can reduce leakage detection time and costs further.

The approach of step-testing using real time monitoring now provides the opportunity to carry out a halving and quartering technique where the zone in question is "sectioned off" for very short periods of time. This helps minimizes the disruption to supply as length of mains are not left in the closed position for a period of time. This is essential in regards to water quality and as society is becoming more active over the night time period.

A valve which will allow the zone to be cut in half is located and closed to see which half contains the leak; this subzone is then halved again and so on until the leak is located to a section of main. There are no criteria as to how this is completed but potentially a leak could be found by only closing a couple of valves.

As with the traditional method of step testing; customer consumption must be estimated for each half, quarter and any subsequent division. The operator can then compare the actual flow rate to the expected flow rate, again any significant difference raises an indication for further leakage investigation (Figure 4.41).

Chapter 5

Case studies

5.1 CASE STUDY: NEW BRAUNFELS UTILITIES (NBU), TEXAS, USA CUTS WATER LOSS BY 50%

Justin Robinson, Marketing Manager, HWM

5.1.1 Abstract

Texas-based New Braunfels Utilities (NBU) used a range of leak detection equipment manufactured by HWM to build an efficient maintenance program for its distribution network while drastically reducing water loss. The equipment has allowed NBU to conduct scheduled repairs on its pipelines instead of dealing with leaks on an emergency basis.

5.1.2 Introduction

The NBU leak detection and valve maintenance program was established in 2009 to reduce water loss and increase system and valve reliability for NBU's 734 kilometres of pipeline and 24 000 customer connections.

5.1.3 Problem description

At the end of the first year of the program, NBU calculated its average water loss at 4.700 litres per kilometre per day.

5.1.4 Solution provided

Recognizing the need for improvement, NBU used HWM leak detection equipment including Xmic ground microphones, SoundSens "i" correlating noise loggers, a MicroCorr Touch leak noise correlator and Permalog acoustic leak noise data loggers.

Permalog data loggers attach magnetically to pipelines and use advanced algorithms to discern the acoustic signature of leaks from background noise. SoundSens "i" and MicroCorr correlators (which are known as TriCorr in the US) analyze data from acoustic sensors to approximate a leak's location. The Xmic electronic ground microphone amplifies noise generated by water escaping from buried supply lines under pressure, allowing users to pinpoint a leak's position.

The team began using this equipment to perform preventive maintenance on 750 valves per year and proactively scan the city for non-surfacing leaks.

5.1.5 Results obtained

After two years, NBU estimates its average water loss at 1760 litres per kilometre per day, which is less than half the loss rate during the program's first year.

"Everyone knows that water is a precious resource and its preservation requires a ton of attention," said NBU Operations and Maintenance Division Manager.

5.2 CASE STUDY: 'LIFT AND SHIFT' LEAK MONITORING REDUCES LOSSES AND COSTS FOR VEOLIA WATER

Justin Robinson, Marketing Manager HWM

5.2.1 Abstract

Veolia Water has been making use of recent developments in water leak detection technology to effectively manage water networks with increased efficiency and at lower cost. HWM's Permalog + noise loggers have been used by the company in a 'lift and shift' method to quickly, easily and accurately detect leaks which their technicians can then repair.

5.2.2 Introduction

Veolia Water wanted to survey, in order to identify leaks, fifteen of their District Metered Areas (DMA), containing approximately 28 000 properties.

5.2.3 Problem description

The speed and effectiveness of the search for leaks in a large area is crucial in minimising waste and reducing Non Revenue Water (NRW) levels because it determines the speed with which any identified leaks can be repaired.

Therefore, the nine technicians involved in the project required a solution that would enable the best use of their time and resources to cover the largest possible area in the shortest time possible to effectively 'audit' the network for leaks. Time lost is water lost.

5.2.4 Solution provided

The company decided to use Permalog + leak noise loggers to identify leak areas within the network. Mounted directly onto the pipes by a strong magnet, and battery-powered, the loggers monitor the section of piping around them for the sound of water escaping under pressure.

The loggers were used in "lift and shift" deployment mode to survey the network quickly and efficiently. This enables operators to deploy loggers, quickly identify potential areas of leakage and then re-deploy the loggers into another area.

Permalog + loggers in "lift and shift" mode are configured to only transmit data when an operative swipes them with a magnet. There is no radio interference from other loggers, and the data is kept both secure and compartmentalised. This is useful when the same loggers will eventually be deployed in many different locations.

Using the system, the operator automatically recorded each deployed logger's serial number and location by GPS tracking. This made deployment very quick, and there was danger of misplacing loggers – or any physical maps – which naturally improved the retrieval time too.

Leakage Teams deployed loggers throughout the targeted area where they remained overnight. The following day, the loggers were collected by the team, and a record made of any leaks identified in a specific location. The loggers were then deployed to a new area. In this way, the team was able to conduct a rolling, sweeping search for leaks quickly throughout the network.

5.2.5 Results obtained

Veolia Water reported excellent results from the operation, finding leaks with great speed and ease, leading to reduced costs from resources. The fifteen DMA's containing approximately 28 000 properties took the nine leakage technicians only 32 days to sweep. In that time 96 leaks were identified.

Each technician, working alone, was able to deploy and collect over 80 loggers a day. The new GPS mapping technology gave the company an audit trail of loggers, allowing them to track the work of each technician and locate the position of each logger precisely. The deployment rate was the highest yet seen, and the speed of the activity enabled detection time to be reduced when compared to other leak-finding technologies and methods.

5.3 CASE STUDY: LEAK NOISE CORRELATOR AND GROUND MICROPHONE TECHNOLOGY USED IN ZIBO CITY, SHANDONG, CHINA TO PINPOINT LEAKS IN THEIR NETWORK

Justin Robinson, Marketing Manager, HWM

5.3.1 Abstract

The Zibo City Water Company in Shandong Province, China, recently used HWM's MicroCorr Touch leak noise correlator to pinpoint a leak on a pipe in a busy urban area.

5.3.2 Introduction

Leak noise correlation is a proven method to pinpoint leaks on underground water pipes. The correlator measures the time difference at which leak noise is received by sensors deployed at either side of the suspected leak. Leak noise travels at a constant velocity, which depends on the material and diameter of the pipe. As a result, the time difference between the arrival time of the leak noise at each sensor, combined with the velocity, enables the location of the leak to be accurately pinpointed – often to within a few centimetres.

The water company knew that there was a substantial underground leak on a section of pipe approximately 140 metres long, made of ductile iron material. The diameter of the pipe was 500 mm and it was located in an urban area.

5.3.3 Problem description

The cost of urban road excavation is very high and so the leak must be precisely pinpointed to carry out the excavation.

Most traditional correlators provide default filter settings for a wide range of different pipe materials and sizes. However, due to the number of unknown variables that can change the frequency of leak noise, default filter settings will never be optimised for every leak situation. Experienced users can manually change the filter settings to find the leak, but this is often a time consuming process involving multiple correlations. Even then it can be "hit or miss" to successfully locate the leak.

5.3.4 Solution provided

The water company used HWM's MicroCorr Touch leak noise correlator which uses accelerometer sensors of unparalleled sensitivity in the marketplace. In addition, the correlator incorporates a unique automated filter intelligence system (AFIS) which automatically runs 55 different filter combinations on the data, checking the quality of the result and optimising the filter settings as required, until the most accurate result can be determined. Working on live or pre-recorded data, tests have shown AFIS to significantly improve leak pinpointing on almost any pipe.

5.3.5 Results obtained

When the MicroCorr Touch was deployed it immediately identified the leak and then determined its position to within 0.1 m. The accuracy of the result was confirmed by using an Xmic ground microphone to listen to the sound of the leak above ground at the precise location specified by the correlator and this was confirmed by the subsequent excavation.

5.4 CASE STUDY: REDUCING LEAKAGE AT THAMES WATER

Antony Green, Vice President, GL Industrial Services

5.4.1 Abstract

Thames Water Utilities Limited (TWUL) is actively managing their water leakage to meet the targets agreed with the Office of Water Services (OFWAT). It is also imperative for TWUL to reduce the existing leakage levels in the most economical way possible. In 2010, TWUL appointed GL to provide leakage consultancy services, to support TWUL in leakage assessment and identification.

5.4.2 Introduction

Following a number of successful projects using GL's burst finding technology within TWULs leakage management program, this project made use of the General Packet Radio Service (GPRS) to transmit data from the pressure loggers to provide increased access to field data and enable increased analysis frequency to identify unexplained anomalies and system leakage. 25 District Metered Areas (DMAs) were evaluated in the North East, Central and West London Zones.

5.4.3 Problem description

For each of the 25 DMAs identified by TWUL, GL were required to:

- Produce a report and presentation for each DMA identifying areas of interest for field investigation and possibly repair works.
- Final reports and presentations detailing the findings and the success rate of the whole project.

5.4.4 Solution provided

DMAs with high leakage or demand irregularities were identified by TWUL and handed to GL for analysis. Data handover includes hydraulic models, GIS maps, and Bow data. Hydrosave, GL's sub-contractors, carried out initial leakage surveys and deployed GPRS pressure loggers at specified locations in the DMA. Pressure data was downloaded remotely via GPRS network using software provided by Halma Water Management.

The hydraulic model was assessed and updated based on the current GIS and Flow data and was then prepared for the "Burstfinder" analysis with the logged pressure data as reference pressures.

"Burstfinder" was run to recalibrate the leakage component of the demand against the current logged pressure data. It provides an 'automated' approach using optimisation to match the chosen variables against the field data.

All results were reviewed and areas of interest assessed to identify anomalies and areas for further investigation in the field. Leakage technicians then carried out leak detection in all areas of interest in the field to confirm the cause of the 'hotspots' for example new demand, leak and so on. All identified leaks were communicated to TWUL for follow-up validation and repair.

5.4.5 Results obtained

- The study results have shown that the "Burstfinder" is highly effective in locating leakage hotspots and significantly improved the performance of leakage teams.
- Through the use of GPRS pressure loggers GL was able to deliver a far more efficient service and provide a full comprehensive set of results against a targeted DMA programme.
- GPRS loggers provided near real-time pressure data, which enabled the network to be monitored continuously and remotely for increased efficiency and faster response time.
- The use of "Burstfinder" enabled GL to proactively direct detection teams to street locations within a DMA, to detect leakage and improve the ESPB rate per FTE and MLD found per FTE
- The use of GPRS loggers allowed repeat analysis to be done instantly after detection or repairs, to allow additional hotspots to be identified within the DMA.
- GL also developed a new form analysis that enabled the identification of demand anomalies and a street location and time period be highlighted for onwards investigation. (Figures 5.1 and 5.2)

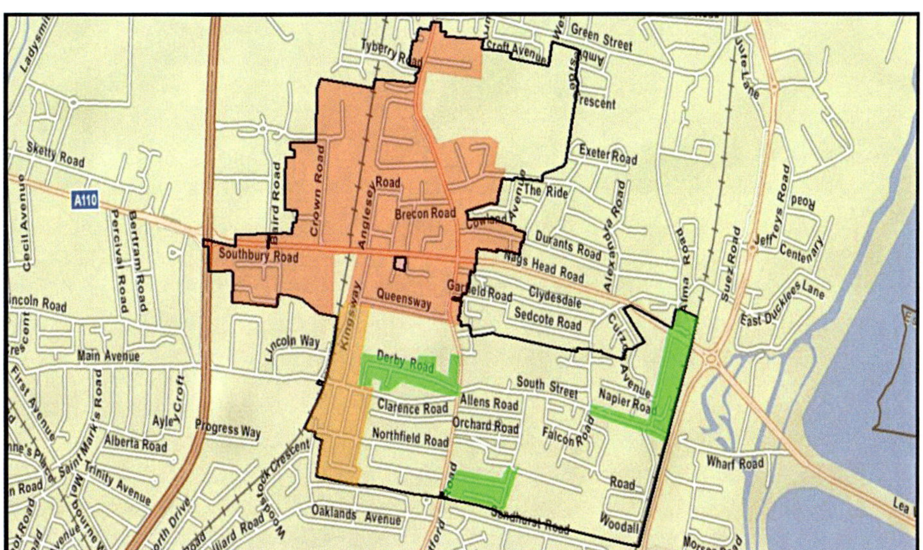

Figure 5.1 Output example from the simulations detailing 'hotspots' in DMA Sewardstone 38 (*Source*: GL Industrial Services).

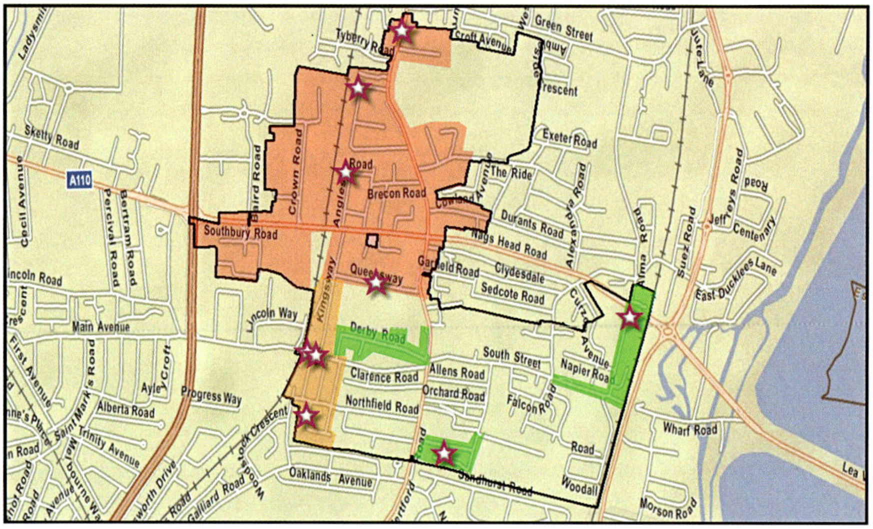

Figure 5.2 Confirmed leaks identified by Burstfinder (*Source*: GL Industrial Services).

5.5 CASE STUDY: LEAK DETECTION FOR ANKARA WATER AND SEWERAGE ADMINISTRATION (ASKI)

Stephen Rothwell, Marketing Specialist, Pure Technologies

5.5.1 Abstract

Ankara Water and Sewerage Administration (ASKI) completed a successful SmartBall® survey of a portion of its large diameter water transmission main that serves the capital city.

5.5.2 Introduction

Ankara Water and Sewerage Administration (ASKI) serves about 4-million customers with water in Ankara, Turkey's capital city, and operates more than 10 000 kilometres of water transmission and distribution mains, including over 700 kilometres of large-diameter mains (500-millimetres and above). The utility wanted to check the condition of one of its large-diameter transmission that is essential in supplying Ankara with water. This particular main, called Ividik-T17, is made of 2200-millimetre Prestressed Concrete Cylinder Pipe (PCCP) and is about 35 years old.

5.5.3 Problem description

Before beginning the leak detection survey, ASKI was struggling with a high rate of water loss per month, with about 36 percent of its pumped water lost. These losses can be attributed to leaks, illegal connections, unbilled water use (fire hydrants etc.) and flushing of the pipeline system. While not all of the 36 percent loss was a result of leaking pipes, the utility was concerned nonetheless and estimated that 80 percent of the losses were a result of leaks. Another consideration for ASKI was the condition of this pipeline; the transmission main was over 30 years old and is a crucial supply of water to its Ankara. Because it has no redundancies that would allow the pipeline to shut down for repair, ASKI wanted to understand the condition of Ividik-T17 to proactively plan contingency measures and potential repairs for one of its major pipelines.

5.5.4 Solution provided

To inspect the transmission main, ASKI used "SmartBall", a free-swimming tool that traverses pressurized pipelines locating and recording acoustic activity associated with leaks and pockets of trapped gas. The tool can be inserted into pipelines of any material while the line remains in service, and travels with the water flow for up to twelve hours, meaning it is able to survey long distances in a single deployment. Because the tool creates no noise, it is able to locate leaks that are very small, and has location accuracy typically within 3-metres or closer to the actual leak. For the 15 kilometre inspection of Ividik-T17, there were 17 SmartBall Receiver (SBR) locations places along the pipeline. This allowed for accurate tracking of the tool as it traversed the pipeline. The SmartBall was inserted at a 150-millimetre gate valve located shortly after a water treatment plant and was extracted at a reservoir. The 15 kilometre "SmartBall" inspection focused on the highest-risk portion of the PCCP pipeline and was completed in November 2011.

5.5.5 Results obtained

The results of the inspection were very positive for ASKI. The SmartBall located 10 total leaks along the Ividik-T17 transmission main, one of which was very large. The large leak (Figure 5.3) was causing a loss of 50 000 cubic metres of treated drinking water per year, with an estimated financial loss of $75 000. This particular leak was caused by a leaking valve beneath the ground, with the lost water finding its way to a river. ASKI was able to fix this leak shortly after its discovery. In addition to the large leak, 9 other leaks were found, 6 of which were located on pipeline features (valves etc.), while only 3 leaks were located on the actual pipeline. The location of most of the leaks on pipeline features was also positive for ASKI, as it showed that the condition of the transmission main was fairly good. Because Ividik-T17 has no redundancies, most of the leak repairs have been postponed until an alternate transmission main is built to supply Ankara, a project that is slated for 2012. The survey helped to establish a baseline condition of the pipeline and reduce water loss.

Figure 5.3 Leak located in excavation (*Source*: Pure Technologies).

5.6 CASE STUDY: LEAK DETECTION PROGRAM IN MANILA, PHILIPPINES

Stephen Rothwell, Marketing Specialist, Pure Technologies

5.6.1 Abstract

Maynilad Water Services Inc. (MWS) in Manila, Philippines, began a comprehensive leak detection program to help the utility deal with their Non-Revenue Water (NRW) problem.

5.6.2 Introduction

MWS is the water and wastewater services provider for 17 cities and municipalities that comprise the West Zone of the greater Metro Manila area, operating about 6000 kilometres of pipeline in the area. The leak detection program began in January 2010 and is an ongoing effort addressing NRW in Manila. The initial contract between Pure Technologies and Maynilad was signed in 2009 as part of a strategic water loss management program aimed at reducing water loss. Under this contract, Maynilad engineers were trained on Sahara® technology to operate the equipment in the West Zone concession. MWS initially rented two Sahara units for ongoing use, but have since reduced to one unit with both audio and visual capability. Pure provides Sahara technical support, maintenance and parts as needed for the utility.

5.6.3 Problem description

Before the beginning of the program, the utility was facing a significant NRW problem, particularly with leaks not visible at the surface, illegal connections, and unknown laterals and old service connections that were unaccounted for. Presumably, many of these problems went unnoticed for a long period of time because they were not visible at ground level, causing significant water loss and fiscal impact.

5.6.4 Solution provided

To address the problem, the client used a tethered leak detection tool, Sahara, which is ideal for urban areas like Manila that have complex interconnecting networks. Pipeline inspections using Sahara are conducted while the main remains in service by inserting a sensor into any tap 2-inches or larger. A small parachute uses the flow of water to draw the sensor through the pipeline. The sensor is tethered to the surface, allowing for real-time results and maximum control and sensitivity as the tool can be pulled back and forth to re-inspect certain areas. The surface tracking device allows the position of leaks and other pipeline features to be located to within 46-centimetres (18-inches). Depending on pipeline features and flow velocity, the inspection distance can be up to 2 kilometres. Sahara Video, which runs concurrently with the leak detection inspection, allows CCTV inspections on pipelines while they remain in service. Sahara video has been used to assess the condition of pipeline interiors, locate known features with unknown locations, and locate unknown features, including unknown laterals and illegal connections along the inspected pipelines.

5.6.5 Results achieved

The leak detection program allowed MWS to establish the condition of their pipeline, which has helped optimize their repair and replacement programs and allowed them to maintain their service to customers. Since the start of the program, 264 kilometres have been inspected with 319 leaks located and 173 illegal connections or unknown laterals identified and shut down. Initially, Maynilad was only doing leak detection using Sahara audio, but have since adopted video as well. A combined audio and video program has helped MWS detect leaks, but also see video anomalies, such as obstructions, protrusions, gas pockets, illegal taps and unknown laterals and valves.

(a)
(b)

Figure 5.4 (a) Illegal connections found on excavation and (b) Staff training in use of equipment (*Source*: Pure Technologies).

5.7 CASE STUDY: LONG DISTANCE LARGE PIPELINE INSPECTION

Fabio Orland, Commercial Director, JD7 Pipeline Services Ltd

5.7.1 Abstract

A large European Water Supply Company, decided to undertake a pilot Pipeline Inspection project to monitor several strategic trunk mains for leakage and condition assessment locateds within its extremely busy town centre.

5.7.2 Introduction

The Company serves approximately 4 300 000 customers through an extensive network of 2 million water meters and a 9 500 km of water pipes.

The Pipeline Inspection programme began in July 2012, and lasted 6 weeks to cover a total distance of 15 km. The programme delivered the client full CCTV Inspection, and also a full leakage sweep of the network.

5.7.3 Problem description

Prior to the project commencement, the clients understanding was that there was not a big issue with leakage due to the ground structure in and around the city, this being very clay rich, in theory would bring all leaking water to the surface. However, due to the importance of the specific mains and the fact that they have been in service for over 30 years without inspection it was decided that an in-line survey will be carried out to establish a baseline for condition assessment and look for any leaks at the same time.

5.7.4 Solution provided

The company decided to use the LDS 1000 which comprises a tri technology head, this being a high resolution colour camera, hydrophone and high powered sonde to give the operator live video and audio feedback during an inspection. The neutrally buoyant cable is then floated within the pipeline utilising the water flow.

The system enters a pipeline via a 2″ tapping, and is fully chlorinated during its Insertion; the system works on a live basis, with no interruptions to the clients services, and can cover a distance of up to 2 km per day.

The technology is the latest trunk main inspection system on the market being fully battery powered and only requires a 2 man team for its implementation.

Figure 5.5 Team operating JD7 LDS1000 in Greece (*Source*: JD7 Ltd).

5.7.5 Results obtained

A total of 15 km of mains were inspected ranging from 400 mm to 700 mm in diameter. A total of 15 substantial leaks were detected, verified and repaired during the project, along with several objects of interest discovered. Images below show one of the Items of interest discovered during the Inspection Project a 2.7 metre long piece of timber.

Figure 5.6 Length of wood located and removed from main (*Source*: JD7 Ltd).

The LDS 1000 was also used to update the clients GIS system by identifying numerous unknown bends, lateral connections and paved over valves, along with validating the condition of existing repairs.

The project was very demanding, as the city is very heavily populated with large volumes of traffic. Due to the LDS 1000 being powered off batteries the system was charged throughout the day, this enabled us on several occasions to work through the night with silent work operation and with far less traffic.

The LDS 1000 has also been successfully used in Canada, Jakarta, Georgia, Mexico, New Zealand and the USA.

5.8 CASE STUDY: POTABLE WATER PIPELINE INSPECTION IN NORTH AMERICA

Fabio Orlandi, Commercial Director, JD7 Pipeline Services Ltd

5.8.1 Abstract

JD7 Pipeline Services were recently mobilised to Canada for a pipeline inspection project to assist a Contractor in locating possible leakage problems in a newly laid pipeline.

5.8.2 Introduction

The Contractor is one of Ontario's largest and specialises in major road, highway, sewer and water main projects and has undertaken a five mile road rehabilitation and water main/hydrant replacement project. The City is one of Canada's fastest growing municipalities, and is located just north of Toronto.

5.8.3 Problem description

The project is significantly behind schedule due to repeatedly failing hydrostatic pressure tests required for certification by the local utility. The pipeline, a 6" PVC C900 pipe with 5 bar pressure using Cobra Locks, was Iinstalled using trenchless directional drilling to reduce disruption to a now established neighborhood.

The contractor outsourced local specialist pipeline inspection companies, and searched for leaks over a four week period using noise correlators and vacuum excavation joints. With the former no leaks were identified whilst with the latter 5 leaks were detected however this was very intrusive and also costly "guess work". The Estimated cost over a 4 week period of failed Inspection/leak detection was very high, about £300 000.

This had a major knock on effect, as the client could not proceed with service hook-ups and road paving until the hydrostatic test is passed on all locations, this also incurred a daily fine issued by the local utility company.

5.8.4 Solution provided

The "Investigator" by JD7 was used to provide an in-line of the pipeline. It has been designed and developed for small distribution mains, sizes 3–12 inch. The system comprises of a tri technology head, this being a high resolution colour camera, hydrophone and high powered sonde, tethered to a 100 m semi rigid rodding to give the operator live video and audio feedback during an inspection. The system enters a pipeline via a 2" tapping, and is fully chlorinated during its Insertion; the system works on a live basis, with no interruptions to the clients services, and can cover a distance of up to 1 km per day. The technology is the latest live main inspection system on the market being fully battery powered and only requires a 2 man team for its deployment.

JD7 Pipeline Services completed the inspections in 10 working days, and in total successfully identified and located 10 leaks all verified when excavated.

Figure 5.7 (a, b) leak located on MDPE electro-fusion joint (c) Located joint on MDPE pipe where leak located (*Source*: JD7 Ltd).

5.8.5 Results obtained

The following results were obtained following the deployment of JD7 Investigator:

- The Investigator™ identified 10 leaks during the project, all were later validated
- Tethered insertion technology system allowed for precise location of the leaks to be identified
- Acoustic system is very sensitive and able to pick up small and large leaks
- Operator was able to identify multiple leaks in close proximately to each other
- The only technology that precisely locates leaks in PVC pipe

The contractor passed the hydrostatic tests at a fraction of the cost and time_they invested until then using various other methods in previous weeks.

The Investigator has also been successfully used in Jakarta, Georgia, Mexico, New Zealand, UK and the USA.

Chapter 6

Paper 1: Water balance – From the desk top to the field

B Charalambous[1] and S Hamilton[2]

[1]*Managing Director, Hydrocontrol Ltd, Lemesos, Cyprus. E-mail: bcharalambous@cytanet.com.cy*
[2]*Hydrotec Ltd, The Barn, Thorpe Underwood, Northamptonshire, NN6 9PA, UK,*
E-mail: shamilton@hydrotec.ltd.uk

Summary

Have you ever wondered what you do with the water balance after the various components have been calculated? Or how you can use the numbers to work out an investment strategy and an action plan and how to prioritise your actions in order to get the best return on your investment? This paper shows how the water balance can be used to derive an action plan for reducing non-revenue water as well as the relevant returns on investment for each action. Also it gives a full working model of how the Water Balance is taken to the next stage.

Furthermore it details the actions that are appropriate to be taken for each one of the above main components of the water balance in order to reduce non-revenue water and provides justification for each action proposed. It also expands on current thinking and knowledge in the planning and prioritisation of non- revenue water reduction options that are available to water utilities and recommends a basic action plan matrix.

Keywords: Water Audit, NRW reduction strategies, NRW action plan matrix

6.1 INTRODUCTION

Most water utilities use the water balance to calculate non-revenue water and to find the amount of water being lost. It is obvious that this is extremely useful and must be worked out in order to have a clear picture and to account for each constituent component of the water balance.

The planning of non-revenue reduction activities to be carried out and ultimately the compilation of an action plan are based on the findings of the water balance and in particular on the main components, namely, Authorised Billed Consumption, Unbilled Authorised Consumption, Apparent Losses and Real Losses. Depending on the amount of water which is being lost in each one of the above components which comprise the Non-Revenue Water, the action plan is targeted in order to provide the best return with the minimum of investment in the shortest time possible.

Accountability of water is extremely important in this process. This is achieved through a validated Water Audit which could be carried out internally by experience water utility personnel or by external auditor.

6.2 WATER AUDIT

A water audit is a thorough accounting of all water into and out of a utility as well as an in-depth record and field examination of the distribution system that carries the water, with the intend to determine the operational efficiency of the system and to identify sources of water loss and revenue loss. It should include but not limited the following:

- A thorough accounting of all water into and out of a distribution system.
- A Water Balance calculation including inspection of system records and data verification.
- A meter testing and calibration program.

A water audit is a critical first step in the establishment of an effective water loss management program. With the successful completion of a system water audit, the utility gains a quantified understanding of the integrity of the distribution system and begin to formulate an economically sound plan to address losses. Water loss in a public water system can be a major operational issue. Non-revenue water components can significantly affect the financial stability of the utility. Addressing the issues associated with the non-revenue components will certainly entail a significant cost for the utility. The economic trade-offs between value of lost water given it generates no revenue and the investment to reduce this loss requires careful planning and economic judgment. The utility needs to clearly understand the type of loss as well as its magnitude. Water resource, financial and operational consequences must be weighed when considering these issues and the decision taken is unique to every system.

A brief summary of the main steps to perform an initial water audit is given below for ease of reference:

(1) The amount of water put into the distribution system is determined.
(2) The authorised consumption (billed + unbilled) is obtained from records.
(3) Water losses are calculated (water losses = system input – authorised consumption).
(4) Apparent losses are estimated (theft + meter error + billing errors and adjustments).
(5) Real losses are calculated (real losses = water losses – apparent losses).

The above steps are an example of a **top down** audit, which starts at the "top" with existing information and records. It may also be known as a desktop audit or paper audit since no additional field work is required. Distribution systems are dynamic. The audit process and water balance has to be periodically performed to be meaningful to a utility's water loss management program.

After performing an initial top down audit it may become evident that some of the numbers are approximate estimates and inspire little confidence in their accuracy. The next action in the audit process is to refine the quantities that may have been initially estimated and begin reducing non-revenue water losses. A **bottom up** approach is often implemented after top down audit has been completed which can help in identifying the real losses component more accurately thus adjusting the initial Apparent losses estimate. A bottom up approach will help with finding real losses and begins by looking at components or discrete areas in the distribution system. It also assesses and verifies the accuracy of the water loss data associated with individual components of the distribution system.

It is important to stress that although utility personnel are well experienced and are familiar with the operational characteristics of the network it may be worth while having an external or independent audit carried out. External audits are usually an excellent way of helping water utilities to analyse and improve their data. It must stressed the external audits are an independent process, ensure accurate reporting, improve data collection and accuracy by identifying statistical and reporting errors and is an excellent method of helping utilities to improve their performance.

There are many types of audits that will analyse water use, from distribution system balances to household reviews. The accuracy of results depends on the methods used to generate the data. Audits have an important part to play in the development of strategic action plans for water efficiencies and financial savings as well as short and long term management. Therefore it is vital they are undertaken in ways which ensure that the most accurate data possible is generated (Queensland Environmental Protection Agency / Wide Bay Water, Manual 2, Water Audits p43).

6.3 ASSESSING LOSSES – IWA WATER BALANCE

A significant contribution to reaching the point of water accountability was the establishment of the IWA Water Balance (Figure 6.1) which is a useful tool in analysing the various components of water production, storage and distribution. Through this analysis the utility will gain an understanding of the magnitude of the water loss problem and will set priorities for rectifying the situation based on the component analysis of the Revenue and Non-Revenue Water elements.

<table>
<tr><td rowspan="9">System Input Volume</td><td rowspan="4">Authorised Consumption</td><td rowspan="2">Billed Authorised Consumption</td><td>Billed Metered Consumption</td><td rowspan="2">Revenue Water</td></tr>
<tr><td>Billed Unmetered Consumption</td></tr>
<tr><td rowspan="2">Unbilled Authorised Consumption</td><td>Unbilled Metered Consumption</td><td rowspan="7">Non Revenue Water</td></tr>
<tr><td>Unbilled Unmetered Consumption</td></tr>
<tr><td rowspan="5">Water Losses</td><td rowspan="2">Apparent Losses</td><td>Unauthorised Consumption</td></tr>
<tr><td>Customer Meter Inaccuracies and Data Handling Errors</td></tr>
<tr><td rowspan="3">Real Losses</td><td>Leakage on Transmission and Distribution Mains</td></tr>
<tr><td>Leakage and Overflows from the Utilities Storage Tanks</td></tr>
<tr><td>Leakage on Service Connections up to the Customer Meter</td></tr>
</table>

Figure 6.1 IWA Water Balance (*Source*: IWA Water Loss Specialist Group).

The findings from the water balance and in particular its main components should:

- Assist in estimating the best return with the minimum of investment in the shortest time possible
- Form the basis for planning NRW reduction activities
- Provide sufficient information for an effective action plan

It is strange however, that for a number of reasons instead of following the above desired results a different approach is adopted which follows the steps below:

- Limited and / or unreliable data is used
- Calculate non-revenue water
- Find the amount of water lost
- Do not like the outcome
- Change assumptions made to suit
- Management 'blaming' staff for not doing their job
- Employees pointing out lack of funding, commitment and support by management
- Finally, work out figures to suit management and employees

Obviously the above approach will result in serious problems for the utility and it must be avoided at any cost. The Water Balance is a useful tool which if used correctly it will certainly point the way to the actions and measures that need to be taken to reduce NRW. Answers to the questions below will take you to the next stage from the desk top to the field environment.

- Have you ever wondered what you do with the water balance after the various components have been calculated?
- Or how you can use the numbers to work out an investment strategy and an action plan?
- Or how to prioritise your actions in order to get the best return on your investment?

Answers to the above questions and what could be done with the water balance will be demonstrated using examples from case studies.

6.4 CASE STUDY EXAMPLES

6.4.1 Top down approach

In this example the constituent components of the water balance are entered into the water balance using absolute volume figures and working out the corresponding percentage figures. It is this percentage figures which are usually quoted and has to be stressed that they could be misleading as a performance indicator since they are strongly influenced by consumption as well as changes in consumption.

The Non-Revenue Water is often expressed as a percentage of the System Input Volume. However, a true financial performance indicator needs to reflect costs as well as volumes. An improved financial indicator can be used by converting the Non-Revenue Water Volume to values. An example is shown in Table 6.1 below where the NRW volumes in the above Water Balance were converted to values using the corresponding unit value for water. The unit value for Unbilled Authorised Consumption and Apparent Losses is usually the average sale price of water to customers. The unit value for Real Losses is usually taken as the marginal cost of water that is the unit cost of producing and distributing water into the network or bulk charge whichever is the higher.

From Table 6.1 it can be seen that the Real Losses have the biggest financial loss for the utility and it is evident that this area is critical and should be examined further. This examination should provide proof that repairing the leaks and savings this amount of water which is being lost makes financial sense for the utility. In order to arrive at this result the following methodology needs to be followed.

Table 6.1 Converting NRW Volume Components to Values.

<table>
<tr><th></th><th>Components of Non- Revenue Water</th><th>Assessed unit value of NRW component</th><th>Assessed total value of NRW component</th><th></th></tr>
<tr><td rowspan="3">Non-Revenue Water 1 768 862 m³</td><td>Unbilled Authorised Consumption 59 928m³</td><td>1,2 € / m³</td><td>€ 71 914</td><td rowspan="3">Assessed total value of Non-Revenue Water € 1 127 436</td></tr>
<tr><td>Apparent Losses 299 639m³</td><td>1,2 € / m³</td><td>€ 359 567</td></tr>
<tr><td>Real Losses 1 409 295m³</td><td>0,8 € / m³</td><td>€ 1 127 436</td></tr>
</table>

From the top down analysis in Figure 6.2 the amount of detectable losses are 1047938 m^3. This figure is equivalent to a Night Line reduction of 1 047 938 m^3/365 days/20 hrs = 144 m^3/hr. Assuming an average leak of the order of 1.6 m^3/hr then the number of equivalent leaks that should be located and repaired is 90. Given that the network length is 345 km it works out that there is on average 1 leak every 3.83 km. Assume a leakage detection team comprises 2 technicians with an average output of 2.5 km per day, 5 day working week and a weekly charge of €5000/week. The average number of leaks found by the team per week is $5 \times 2.5/3.83 = 3.26$, say 3 leaks found per week. The total time required to find all leaks will be 90 leaks/3 = 30 weeks. Based on the above the following financial calculations can be made

- Total Cost for locating leaks = 30 × €5000 = €150 000
- Total Cost for repairing leaks = 90 × €1500 = €135 000
- Water Saving = 1 047 938 m^3 × €0.8/m^3 = €838 350
- NET SAVING = €838 350 − €150 000 − €135 000 = €553 350

It obvious from the above calculation that the utility will have a considerable saving by moving forward with repairing the leaks first and an action plan to this effect should be work out.

In order to highlight a different approach to the above the apparent losses in Table 6.1 are increased with the corresponding reduction in the real losses. The revised figures are shown in Table 6.2 below.

In this instance the action plan needs to be different to the above for the following reasons:

- The Apparent Losses are almost equal to Real Losses in terms of revenue loss
- Need to deploy a strategy that will maximise benefits
- Tackle apparent losses with the minimum expenditure; reduce unauthorised consumption, meter reading and accounting errors at the first instance which will increase revenue.
- Simultaneously reduce leakage in order to save money in producing / buying less water.
- Invest savings in further reducing Apparent and Real Losses.

<table>
<tr><td rowspan="11">System
Input
Volume
11.985.560
100,00%</td><td rowspan="4">Authorised
Consumption
10.276.626
85,74%</td><td rowspan="2">Billed Authorised
Consumption
10.216.698
85,24%</td><td>Billed metered consumption
(including water exported)
10.216.698(85,24%)</td><td rowspan="2">Revenue water
10.216.698
85,24%</td></tr>
<tr><td>Billed unmetered consumption
Zero</td></tr>
<tr><td rowspan="2">Unbilled
Authorised
Consumption
59.928
0,50%</td><td>Unbilled unmetered consumption
Zero</td><td rowspan="9">Non-revenue water
1.768.862
14.76%</td></tr>
<tr><td>Unbilled unmetered consumption
59.928 (0,50%)</td></tr>
<tr><td rowspan="7">Water Losses
1.708.934
14,26%</td><td rowspan="2">Apparent Losses
299.639
2,50%</td><td>Unauthorised use
59.928 (0,50%)</td></tr>
<tr><td>Metering inaccuracies
239.711 (2,00%)</td></tr>
<tr><td rowspan="5">Real Losses
1.409.295
11,76%</td><td>Real losses on raw water mains and at the treatment works
Zero</td></tr>
<tr><td>Leakage on transmission and/or distribution mains
80.458 (0,67%)</td></tr>
<tr><td>Leakage and overflows at storage tanks
11.986 (0,10%)</td></tr>
<tr><td>Leakage on service connections up to the metering point
268.913 (2,24%)</td></tr>
<tr><td>Detectable Losses
1.047.938 (8,74%)</td></tr>
</table>

Figure 6.2 Top down approach using the IWA Water Balance.

Table 6.2 Converting NRW Volume Components to Values (Revised).

<table>
<tr><td rowspan="4">Non-
Revenue
Water
1 768 862 m³
(14,76%)</td><th>Components of Non- Revenue Water</th><th>Assessed unit value of NRW component</th><th>Assessed total value of NRW component</th><td rowspan="4">Assessed total value of Non-Revenue Water
€ 1 678 917</td></tr>
<tr><td>Unbilled Authorised Consumption
59 928m³
(0,50%)</td><td>1,2 € / m³</td><td>€ 71 914</td></tr>
<tr><td>Apparent Losses
599 639m³
(5,00%)</td><td>1,2 € / m³</td><td>€719 567</td></tr>
<tr><td>Real Losses
1 109 295m³
(9,26%)</td><td>0,8 € / m³</td><td>€ 887 436</td></tr>
</table>

6.4.2 Bottom up audit – case study to show bottom up and top down comparisons

It this example it is explained how the bottom up audit is extremely useful in complementing the top down approach. The case study data used are of the area of 'Sky' in Piraeus, Greece (Kanellopoulou, S., 2011). The main characteristics of the area are as follows:

- Length of network = 56km
- Number of consumers = 16840
- Service connections = 4000

A top down approach is carried out and the result is given in the table below. It should be noted that under the Real Losses the two main constituent components are included:

- Background Leakage on mains and service connections, and
- Detactable Losses

In order to verify the assumption made for the Apparent Losses a bottom up audit will be carried out based on measured values. The Minimum Night Flow (MNF) as measured is shown in Figure 6.3. The Minimum Night Flow is 55 lit/sec (198 m^3/hr). Based on this measured figure a bottom up audit is carried out of the constituent components of the MNF in order to arrive at the amount of potentially detectable losses as shown in Table 6.4.

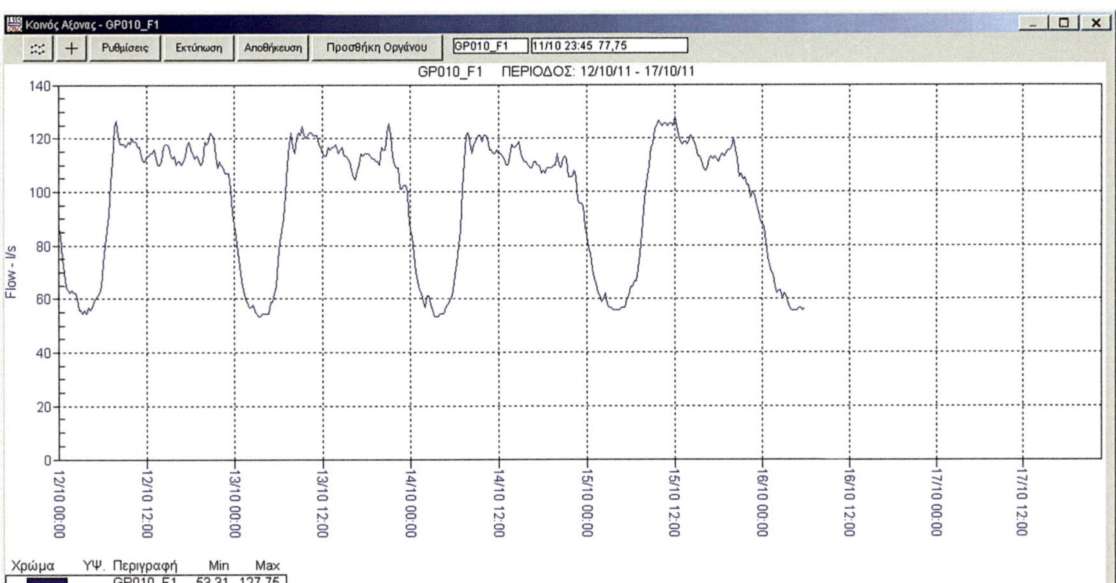

Figure 6.3 Minimum Night Flow Diagram for the area of Sky, Piraeus, Greece. Software: Intranet telemetry application, 2006, EYDAP, Athens, Greece.

As it can be seen from Table 6.4 below the potentially Detectable Losses are 2352 m^3 per day compared to the overall figure of 2631 m^3 per day estimated for all Real Losses in the system using the Top Down approach (Table 6.3).

Table 6.3 Top down approach for the area of Sky, Piraeus, Greece.

Description	M^3/year	Average Daily Volume (m^3)
Input volume	2 898 100	7940
Construction	1 881 575	5155
Non-revenue water	1 016 525	2785
Unbilled Authorised Consumption (measured)		26
Apparent Losses (assumed 2.5% of consumption)		129
Real Losses		**2630**

Table 6.4 Bottom up audit for the area of Sky, Piraeus, Greece.

Description	m^3/hr	Daily (m^3)
Minimum Night Flow (measured)	198	198 × 20hr = 3960
Background Losses (calculated)	21	21 × 24 = 504
Customer Night Use	46	46 × 24 = 1104
Potentially Detectable Losses	3960-504-1104	**2352**

This exercise shows that the initial assumption for Apparent Losses is reasonable and the NRW reduction plan should concentrate on reducing the Real Losses particularly in locating and repairing the leaks in the distribution network. The potentially detectable losses comprise 89% of the overall Real Losses which is of the right order considering that apart from a small percentage of Background Losses and Customer Night Use the remainder is attributed to losses in the distribution network which could potentially be located and repaired. Of course the Economic Level of Leakage must be taken into consideration in deciding how much of the amount of potentially detectable losses is financially worthwhile recovering.

6.4.3 Benchmarking of non-revenue water

It is extremely useful to have a matrix which could be used to benchmark the performance of a utility based on the NRW figures.

The authors have developed and are proposing for use an action plan matrix which is based on the percentage of System Input Volume for each constituent component of the Non-Revenue Water. The action plan matrix which is shown in Table 6.5 provides guidance as to the general actions that could be taken depending on the percentage figure in order to reduce the NRW in each component.

Of course the proposed matrix is only a guideline and much more investigation and development of this Matrix is required. The intention is to provide a guideline as to the general actions required which could be carried out by the utility whilst collecting and validating further data and information for more detailed analysis which will result in specific water loss management strategies.

Table 6.5 Proposed Action Plan Matrix for NRW.

Water Balance Component	% of System Input Volume	Suggested Action
Unbilled Authorised Consumption	Up to 1%	Considered within acceptable limits
	1% to 5%	Introduce new tariffs
	5% and above	Review overall billing policy
Apparent Losses	Up to 2%	Considered within acceptable limits
	2% to 5%	Reduce unauthorised consumption, meter reading and accounting errors
	5% and above	Review metering accuracy / policy
Real Losses	Up to 5%	Considered acceptable, may be uneconomic to reduce
	5% to 10%	Reduce visual leakages and overflows at storages and fix visual network leaks
	10% and above	Improve active leakage control, effective maintenance, pressure management

6.5 CONCLUSIONS

It could not be stressed enough that utilities must target their actions and investments in order to get the maximum benefit. To achieve this it is important to have the necessary knowhow either internally or externally in order to be in a position to justify a proposed NRW reduction action plan which above all should be financially viable and sustainable. Needles to say in order to carry out such a plan the right level of knowledge and experience are required.

So, tackle first whatever gives you the quickest revenue return which will provide money for the longer term savings – think of the returns and not get caught up in the expensive solutions because it may be more attractive.

It is important to be understood that the Water Balance is the starting point for any NRW work. This paper aims to show that this could be done at the very early stages without having to wait until DMAs, pressure management, and so on. are set up and data collected and analysed.

The Water Balance provides sufficient information to assist in the drafting of a NRW master plan in order to move ahead with water loss reduction and in parallel to make strategic improvements to the network. In the past it was thought necessary, mostly in developing countries, to develop DMAs in order to drive a NRW reduction master plan. The authors are suggesting that this could be done in parallel and that in the initial stages the Water Balance is the vehicle for driving a NRW Master Plan.

FURTHER READING

Lambert A. (2003). Assessing Non-Revenue water and its components, *Water21*, **5**(4), 50–51.

Queensland Environmental Protection Agency & Wide Bay Water (2002). Managing and reducing losses from water distribution systems. Manual 2, Water audits, Brisbane, Australia, p9, p43, p54.

Chapter 7

Paper 2: Intermittent supply leakage nexus

B. Charalambous

Managing Director, Hydrocontrol Ltd, Lemesos, Cyprus. E-mail: bcharalambous@cytanet.com.cy

Abstract

The world's population is increasing at a tremendous rate, the world's renewable water resources are reducing rapidly, the gap between supply and demand is widening with urbanisation and climate change making it even wider. This paper reviews how the Water Board of Lemesos, Cyprus, a water utility with a proven history of managing extremely well its distribution network, was forced due to water shortage conditions to have intermittent supply, providing water 3 times a week for about 12 hours each time. Although intermittent supply is perhaps the last measure to be taken in conditions of water shortage, it is however a situation worthwhile avoiding through proactive planning and timely response to critical conditions. The adverse effects of intermittent supply on customer service and on the integrity of the distribution network, increase in number of bursts and leakage, as well as the financial repercussions to the utility are also discussed based on the experiences gained from the intermittent supply measures taken during the water shortage periods faced by the Water Board in the last twenty years.

Keywords: Water shortage, intermittent supply, leakage

7.1 INTRODUCTION

A drought can be defined as a prolonged period of unusually dry weather in an area where some rain might normally be expected. Droughts involve inter alia, water shortages, crop damage, stream flow reduction and depletion of groundwater and soil moisture. A drought happens when a period of low rainfall leads to a shortage of water. It is starting when total rainfall is well below average for several months. A balance must be maintained between the water taken out for supply and that being replaced by surface run-off. Normally the surface run-off during the winter far exceeds demand for supply, so that the excess water can be stored and used when surface run-off is less than demand from consumers, normally during the summer period.

Usually water shortage is declared when the water supplies fall short of meeting water demands. This situation usually prevails in arid and semi-arid areas of our planet where precipitation has been steadily declining or where management of limited water resources has been wasteful and unwise.

Intermittent water supply may be defined as a piped water supply service that delivers water to users for less than 24 hours in one day. It is a type of service that, although little found in developed countries, is very common in developing countries. In an intermittent supply situation the consumers secure their water supply through the use of ground or roof tanks, where water is stored during the length of time that the supply is provided.

It is worth noting that intermittent water supply is enforced not only in cases where there is water shortage but also where the hydraulic capacity of a network is such that can not satisfy demand as well as in cases where the network is severely deteriorated.

In many instances there is no indication how long intermittent supply measures will be in place. The hydrological conditions in each case could impact adversely on water supply for years in which case conserving as much as possible the limited water resources may not be the long term solutions but it may be necessary to add to the water balance new non conventional water resources. In many countries water shortage problems were overcome through the desalination of brackish or saline water. Of course exploring every potential water source available may be the only solution in many instances, but conservation and leakage reduction is always one of the least expensive and quickest solutions to ensuring that water will be available when needed.

7.2 WATER RESOURCES AT GREAT RISK

The scarcity of water resources is one of the most pervasive natural resource allocation problems facing water users and policy makers. This problem is faced each day in the many conflicts that surround the use of this limited resource and the provision of water to the many user groups is determined based on the specific condition pertaining at any point in time.

The pressures that exist on water resources are highlighted by the water stress indicator which measures the proportion of water withdrawal in relation to total renewable resources. As it can be seen from the map in Figure 7.1 a large proportion of the densely populated part of the planet has a high to very high water stress indicator. It is therefore imperative to develop appropriate water management approaches in order to manage our water resources efficiently and effectively.

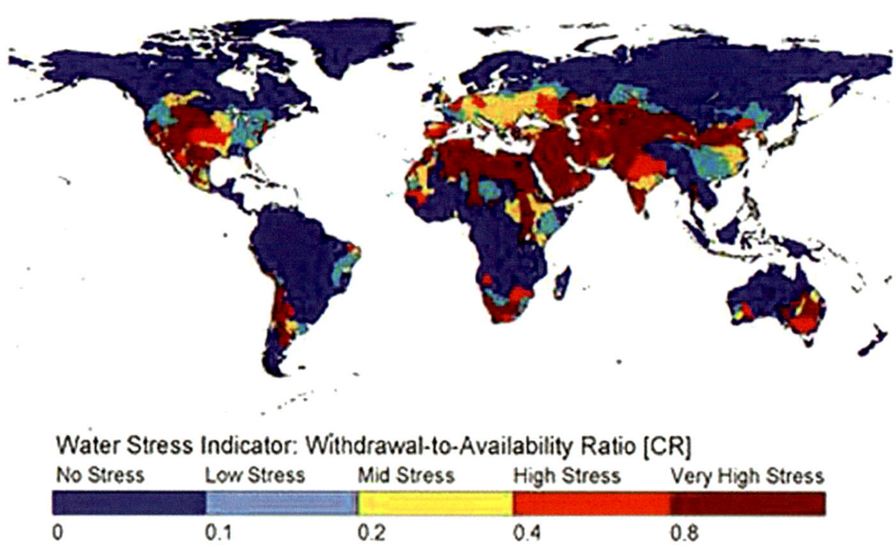

Figure 7.1 Water Stress Indicator: (*Source*: Water GAP 2.0 – December 1999).

Climate change adds to these concerns. It has been affecting the average weather patterns that we were all used to and engineers and scientists have to take this into consideration in present and future planning. As an example, in Cyprus, the largest island in the eastern Mediterranean, the precipitation records of the last 100 years indicate an overall decrease in the mean annual precipitation of about 15%, but annual variation in precipitation varies considerably from the mean with long periods below average, affecting significantly the annual water resources of the island. This pattern is very similar across the Mediterranean basin and there are cases in recent years where cities were even forced to ship water from other countries in order to combat the crisis. For instance, the town of Lemesos in Cyprus was supplied daily by tankers with water from Athens in Greece for an eight month period in 2008/2009 to overcome a serious water shortage problem caused by prolonged drought. During the same period Barcelona in Spain was also being supplied water via tankers in order to relieve a similar water crisis. This phenomenon seems to be growing to global dimensions.

Faced with such pressures, there is the prospect that water authorities will increasingly wish to resort to delivering intermittent supplies. Usually during drought periods water authorities impose water restrictions to both domestic and agricultural supplies. At the same time they move forward with the construction of treatment units to treat domestic effluent for agriculture, and if this measure is not sufficient they resort to the construction of desalination plants to produce potable water for satisfying domestic needs, thus adding to the water balance and reducing deficit. However, in most cases water authorities seem to overlook the obvious, which is to manage the water networks in the most efficient and effective way in order to minimise losses.

7.3 WATER LOSS MINIMISATION

Reducing losses from distribution networks is of the utmost importance and water utilities must recognise this and respond positively. Efficient and effective water loss control should be recognised as a first priority for improving potable water supply.

Accounting for water is the first step that must be made by any utility. It is imperative that an accurate and comprehensive metering system is in place for registering all water along the chain from production to the consumer, including measurement of the water produced and/or imported, water flow in and out of treatment plants and storage reservoirs and into the zones and district metered areas. It is imperative to eliminate or minimise Authorised Un-Metered Consumption thus achieving the highest possible accuracy in accounting for all water produced. Apparent and Real losses must be analysed and action taken as necessary to reduce the Apparent Loss component thus increasing the utility revenues as well as applying the most cost effective leakage programme which reduces leakage to an economically, environmentally and socially acceptable level.

Decision makers at all levels in water utilities must understand that any water loss control strategy, in order to be effective, must be a continuous activity based on a long-term strategy and should form an integral part of the utility's vision. The success of the strategy will inevitably depend on the commitment and dedication at all levels within the utility and of course on the adoption of appropriate strategies and techniques. A successful strategy is one that maintains the distribution network in a proper working order, reducing and maintaining leakage at an economic level, and of course providing the required level of service to all consumers.

7.4 THE WATER BOARD OF LEMESOS CASE STUDY

7.4.1 The distribution network

The Water Board of Lemesos is a non-profit, semi-government organisation charged with the responsibility of supplying potable water to the town and environs of Lemesos. The main activities of the Water Board are: planning and execution of technical projects, operation and maintenance of the water production and water supply systems and all associated financial services including collection of water revenues and determination of water tariffs.

The development of the distribution network took place in an organised fashion with new areas of supply being incorporated into their respective pressure zones, strictly governed by the areas ground contours. Each pressure zone is subdivided into DMA's, a total of 60 are currently operating, which have a single metered source with physical discontinuity of pipe work at boundaries.

In late 1980 the Water Board embarked on a detailed programme of leakage management. Since 2002 the Water Board has adopted the practices and methodologies advocated by the IWA Water Loss Group. The efforts made and importance placed by the Water Board for proper leakage management is reflected in the reduction of the non-revenue water over the years and in the improved operational performance of its network. In 2007 the Board had an ILI below 2 and real losses under 100 litres/connection/day.

7.4.2 Water supply conditions

The Water Board realised that water conservation is not to be equated with temporary restrictions on customer water use. Although water restrictions can be a useful emergency tool for drought management or water shortage situation, water conservation programs concentrate on continuous improvements in water use efficiency. To this end the Water Board embarked on a promotional campaign through television, radio and leaflets to increase public awareness for water conservation.

In 1991, the government legislated against the use of hosepipe for washing cars and pavements at all times, a law, which the Water Board enforces during drought periods only. During the drought period of 1997–2000 the government was forced to announce in early 1997 a reduction of 20% for potable water and 40% for irrigation water supplies. In 1998 the water situation became worse and the restriction measures became more stringent as the available quantities of water were diminishing and the government, much to the discontent of the public, went ahead with further measures, enforcing greater restrictions to the water supplies with targeted figures of 28% for potable water and 56% for irrigation use.

The Water Board of Lemesos always responded promptly to the government's declared drought measures and imposed each time restrictions to the continuous supply. In February 1997 restricted supply to consumers to four days a week. In 1998 due to limited water reserves the Water Board was forced to decrease further the availability of water reducing the time of the water being available to the consumers to 12 hours every 48 hours. These measures were in force until the end of 2000. These actions resulted in an overall reduction in the use of domestic water of approximately 15% per annum which proved that the domestic demand is near enough inelastic.

It was thought that after the above 4 year period of intermittent supply, measures would have been taken so that there will be sufficiency of water quantities for at least the domestic needs. However, this was not the case and 8 years later, in 2008, water cuts for domestic use were imposed once again. Water was brought with tanker boats from Athens, Greece. For many this dire situation is blamed on the politicians for not taking the right decision at the right time. For others, there is the argument that there has been mismanagement of the water reserves prior to the drought period and not sufficient forethought has gone into the planning of such a severe drought taking place so soon after the last one. In any case the fact remains that the Water Board of Lemesos even though it has improved its network to such an extent that it is consider amongst the world's best with losses from the network being extremely low, it was faced with a situation beyond its control. Of course the situation would have been a lot worse if the

Board did not continuously improve its network in order to minimise losses thus saving valuable quantities of water.

Intermittent supply therefore, may have seemed to be the short term answer to the water shortage situations faced in the last 20 years, however in the following sections the adverse effects on the integrity of the distribution network of such actions are discussed and quantified based on the experiences gained.

7.4.3 Effects of intermittent supply

Analysis of case study data showed that there was a large increase in the number of reported pipe breaks during the period of intermittent supply. In order to quantify these a comparison was made for a large number of District Metered Areas, 20 in total, between the breaks reported in 2007, before the intermittent supply was applied, and those reported in 2010, the first year immediately after the measures were lifted. The results are shown in Table 7.1 covering both mains and service connections.

Table 7.1 Effect of intermittent supply on reported pipe bursts.

Description	Number of reported breaks		
	Before	After	% Increase
Mains	1 in 7,14 km	1 in 2,38 km	300
Service Connections	15,5 in 1000	29,7 in 1000	200

This comparison showed that the number of breaks on mains increased from an average of 1 in 7,14 km of mains to 1 in 2,38 km of mains, an increase of 300%. Similarly the number of reported service connection breaks increased from an average of 15,5 in 1000 connections to an average of 29,7 in 1000 connections an increase of approximately 200%.

Of course, in addition to the reported breaks, there is a significant number of breaks, caused by the frequent emptying and refilling of the network, which do not come to the surface since the network is not pressurised for any significant length of time to force the water to come to the surface or to have the opportunity to locate these through active leakage control.

The Minimum Night Flow has increased in all District Metered Areas, a typical example being District Metered Area (DMA) 129. Figure 7.2 shows the Minimum Night Flow before and after intermittent supply for this DMA. From the graph it is evident that leakage in this DMA increased from 10 m^3/hr to 20 m3/hr.

This situation is typical of all DMAs with relatively old, about 40 years, network. In order to reduce leakage the Water Board applied a program of Active Leakage Control commencing with Districts with the highest increase in the equivalent number of pipe bursts.

Calculations have shown that the overall increase in leakage due to the intermittent supply measures was of the order of 9%. This figure was substantiated in an analysis using a 'top down' and 'bottom up' approach for 2007, before the intermittent supply was applied, and for 2010, the first year immediately after the measures were lifted.

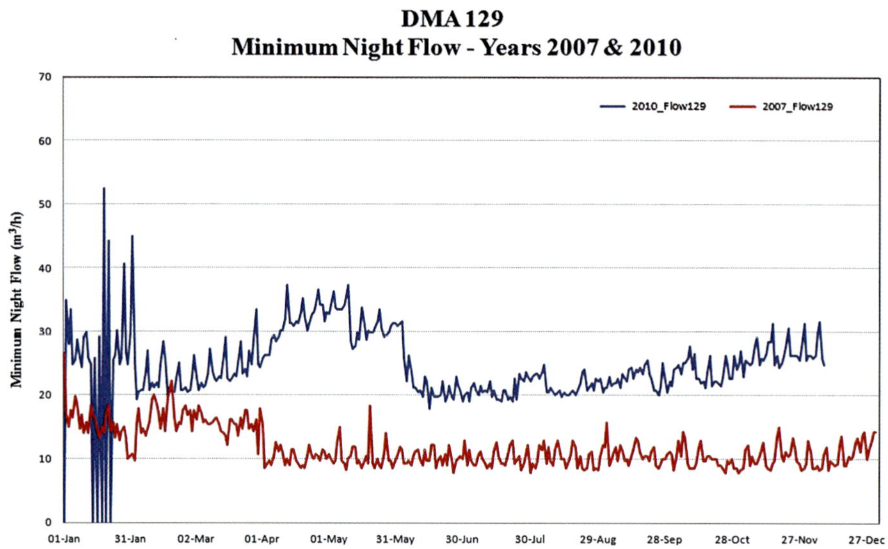

Figure 7.2 Minimum Night Flow Graph for DMA 129.

Figure 7.3 shows the total Minimum Night Flow before (blue colour) and after (red colour) the intermittent supply. It is obvious that there has been a significant increase in the Minimum Night Flow which could only be attributed to the additional breaks which the network suffered during the intermittent supply period. These non reported breaks will have to be located through active leakage control activities and repaired in order to reduce the level of leakage to the level prior to the application of intermittent supply measures.

1 January to 30 December				
	2007	2010	Difference	
Zone 1	5,185,434	5,808,271	622,837	
Zone 2	5,150,320	5,497,431	347,111	
Zone 3	939,659	1,073,681	134,022	
Z4 Ag.Phila	180,690	152,175	-28,515	
Zone 5	141,591	114,837	-26,754	
Zone 6	83,118	134,086	50,968	
Zone 7	20,759	21,509	750	
Total (m^3)	11,701,571	12,801,990	1,100,419	9.40%
MNF (aver.)	385	549	164	m^3/h (period)
Consumers	60,730	66,684	5,954	
Night use	103	113	10	m^3/h
	392,419	430,892	38,473	m^3 (period)
Background losses & Bursts	5,914	9,149	3,235	m^3 (per day)
	1,969,454	3,046,663	1,077,210	m^3 (period)
	16.83%	23.80%	6.97%	
Night use + Background losses & Bursts	2,361,873	3,477,555	1,115,683	m^3 (period)

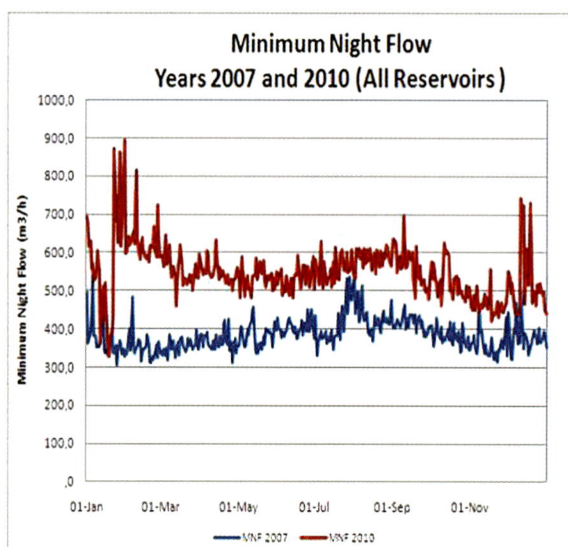

Figure 7.3 Minimum Night Flow Graph for all reservoirs.

From the graph in Figure 7.3 it can clearly be seen that the difference in the Minimum Night Flow between 2010 and 2007 is the additional volume of water being lost due to new breaks caused by the intermittent supply operation. It is calculated that this additional volume is approximately 1 655 000 m^3/year which is slightly greater than the average annual volume of water saved (1 607 000 m^3) due to the intermittent supply measures during the 2 year period of the intermittent supply.

Further evidence from the case study to substantiate the increase in leakage due to the intermittent supply measures is given in Table 7.2 which shows an increase of 12,8% in the System Input Volume after intermittent supply compared with before without a corresponding increase in customer consumption. In fact the customer consumption was slightly less (1,2%) than that of the year before the intermittent supply measures were applied. It is also evident from Table 7.2 that the System Input Volume in the first year of Intermittent supply decreased by 17,5% whereas in the second year by 9,1% indicating that the breaks in the network were increasing in number. This is substantiated by the fact that the reduction in the customer consumption remained effectively the same for both years of intermittent supply.

Table 7.2 System input volume vs customer consumption.

Year	System input volume	Customer consumption
2007 Before Intermittent Supply	Base line 0%	Base line 0%
2008 Intermittent Supply	−17,5%	−9,2%
2009 Intermittent Supply	−9,1%	−8,9%
2010 After Intermittent Supply	+12,8%	−1,2%

Furthermore numerous complaints were received from disgruntled consumers regarding quality problems and of course lack of pressure during intermittent supply. Needless to say that intermittent supply caused serious disruption and upheaval to the daily activities of people whether these were at home or at work.

7.4.4 Cost of intermittent supply

The implementation of intermittent supply has a direct financial cost to the water utility in addition to the loss of revenue due to the decrease in the sales of water.

The loss of revenue to the utility is a direct result of the conservation of water by the customers. Of course, the corresponding system volume is also less which means that overall financial loss will be the difference between the selling and buying/producing the water. The cost of water which is lost through the additional leakage caused by the intermittent supply operation depends on the running time of each leak and the cost of water. It is however a major cost to the utility and one which will continue to burden the utility until the additional leaks are found and successfully repaired.

The direct costs include, amongst other additional operational costs for opening and closing sluice valves to implement water rationing the cost for repairing reported breaks caused to the network due to the frequent emptying and filling of the pipes.

A major financial burden to the utility will be the cost of locating and repairing unreported breaks through an intensive and concentrated effort in order to minimise the running time of the additional leaks.

The Board estimated that the overall average loss of revenue due to the reduction in water sales was of the order of €300 000 for the 2 years of intermittent supply. The additional expenses paid in overtime to staff for the same period of the 2 years was estimated at €365 000 where as the cost of repairing the additional reported breaks was of the order of €325 000.

It is also estimated that the volume of water which is lost due to the increase in leakage when continuous supply was established was estimated to be approximately 1 655 000 m^3 per year which is equivalent to €1 325 000.

7.5 CONCLUSIONS

It is evident from the results presented in this paper that although intermittent water supply may seem to be a solution to a water shortage situation in overall terms the water balance is adversely affected. Supplying less quantity in an intermittent manner causes such deterioration to the network that when continuous supply is re-established additional quantities are lost through increased leakage, which in fact places an added financial burden on the utility.

It is therefore evident that no matter how good a network is, intermittent supply operation has definitely a detrimental effect on its integrity and in addition the amount of water 'saved' is later 'lost' and in greater quantities through increased levels of leakage. Such operational conditions should be avoided especially in pipeline networks that have been designed for continuous supply.

In addition it has been shown that the domestic demand is in effect inelastic and in fact the quantities of water saved by the customers were very small. It is the authors opinion that perhaps better results could be achieved through a structured conservation programme rather than intermittent supply. Of course such programmes are to be introduced as part of an overall strategy for water conservation both on the supply and demand side.

FURTHER READING

Charalambous B. (2007). Effective pressure management of district metered areas. Water Loss 2007 Conference Proceedings, 23–26 September, Bucharest, Romania, Volume **I**, pp. 241–255.

Charalambous B. (2009). Water crisis – bridging the gap. Water Loss 2009 Conference Proceedings, 26–30 April, Cape Town, South Africa, pp. 241–246.

World Water Council (2008). www.worldwatercouncil.org.

Chapter 8

Paper 3: The problem of leakage detection on large diameter mains

Stuart Hamilton[*] and Daniel Krywyj[**]

[*]*Hydro Tec Ltd, The Barn, Thorpe Underwood, Northampton, NN6 9PA, UK*
E-mail: Shamilton@hydrotec.ltd.uk
[**]*JD7 Ltd, Unit 8, Masons Business Park Nottingham Rd, Derby, DE21 6YZ, UK*
E-mail: Daniel.Krywyj@jd7.co.uk

Keywords: Acoustic, Leaks, Non-Tethered devices, CCTV, Hz

8.1 INTRODUCTION

The authors of this paper are communicating the issues with transmission of noise created by a leak on larger diameter water mains and the issues involved with listening to such noise created by the leaks. The authors are also looking at the average frequency of leaks per km and the average losses from such leaks on large diameter mains.

Internal noise frequency from leaks on large diameter mains (greater than 500 mm) in some instances 50 m from the leak position can be as low as 1 Hz–10 Hz, a frequency that no human being regardless of sex or age can detect. For this reason the technology and associated software used today should be such as to identify if a leak is present and not merely relying if a noise can be heard or not. It has always been considered that if a noise cannot be heard then there would be no leak present. In the past this approach has been adopted using conventional and advanced acoustic leakage detection equipment. A modern internal tethered device has been adapted to do leakage detection in large diameter mains by showing CCTV, noise amplitude and frequency and yet even with this information the software graph is ignored if no sound is heard by the technician.

Acoustic noise generated at the point of a leak, regardless of pipe diameter or material, can be higher than 500 Hz, however these higher frequency noises are lost through the pipe wall and the water of the large diameter mains over distance leaving only the low frequencies 1–10 hz – it is these lower frequencies generally below the human threshold of hearing that can travel long distances in the pipeline, sometimes several km. It is for this reason that listening for noise on fittings or correlating is not always successful. Due to these circumstances the internal leakage approach, although may be more expensive, will have a higher degree of success in locating leaks on the larger diameter mains.

Other technology is also available today to indentify these extremely low frequency noises, however, some of this technology cannot locate the leak position but only identify if a leak is present or not.

Manufacturers are also seeking to find the solutions through correlators and to date some successes have been reported.

The following section explains the importance of understanding the parameters involved in conducting leak detection on large diameter mains.

8.2 BACKGROUND

8.2.1 Human ear frequency range

The quality measurement for any noise heard is reliant on how well this noise sounds to the human ear. The human ear is extremely sensitive to noise and high volumes of noise can damage the range of hearing. This is reflected in any time loud noises are listen to, as a ringing noise may be heard afterwards. This is an indication that some damage has been caused.

Hertz (Hz) is used as the measurement of the frequency of sound. For example, a low frequency sound, maybe something that is emitted from an instrument such as a large bass drum, whereas a high frequency sound would be a sound such as that which comes from a whistle. The Hertz sound range is different to how loud or how quiet the noise is, for instance, a load piece of machinery or a whisper.

The human ear has an average hearing range from a low of 20 Hertz (20 Hz) to a high of 20 000 Hertz (20 kHz); but it is the most sensitive to sounds that sit between the range of 1 kHz and 4 kHz. When a person ages their hearing starts to deteriorate and by the time they reach the age of eight, the deterioration has already started. For those that during the teenage years listened to loud music or went to games with loud noises, by the age of 20 they may have already be unable to hear the high tones. Thus the hearing ability has dropped 20 000 Hertz (20 KHz) to 16 000 Hertz (16 KHz). Then by the time the age of 30 is reached, it can be such that the ability to hear the high frequencies has dropped again.

Infrasound is the term used to describe any sound less than 20Hz (i.e below the human range of hearing). Sounds above 20kHz are known as ultrasound (i.e above the human range of hearing).

Animals as well as humans, have the ability to hear over a range of frequency.

- Dogs can sense frequencies as low as 50 Hz and as high as 45 000 Hz.
- Cats can hear frequencies as low as 45 Hz and as high as 85 000 Hz.
- Bats, as nocturnal creatures, need to rely on sound echolocation for navigation and hunting. They can pick up frequencies as high as 120 000 Hz.
- Dolphins are extremely sensitive to frequencies and can sense them as high as 200 000 Hz.
- Elephants, have an unusual ability to detect infrasound. They have an audible range of approximately 5 Hz to 10 000 Hz.

Figure 8.1 is an example of the difference between high and low frequencies waves.

8.3 CASE STUDIES

The data shown in Tables 8.1 and 8.2 is from case studies obtained from water companies from around the world and based on validated findings of unreported leaks. These studies show where these unreported leaks have been located at different pressures, diameters and in various materials. These case studies will indicate the type of leaks located in each material along with associated noise levels. Also reported in this section will be the number of leaks per km and the average size of the leak in m^3/hour

It can be said that although many leaks have been located, not all were repaired as some were beyond economical repair, however these leaks should be regularly inspected as they are all potential catastrophic failures. To date there has not been sufficient data worked upon to identify the life of a leak and it is not known if the burst frequency/natural rate of rise calculation used in the distribution mains is applicable to the larger diameter mains however the authors suggest that recording the date of the survey and the time period measured between surveys can be used as base line data. This data against the age of the pipe will be an indicator to be used in future analysis.

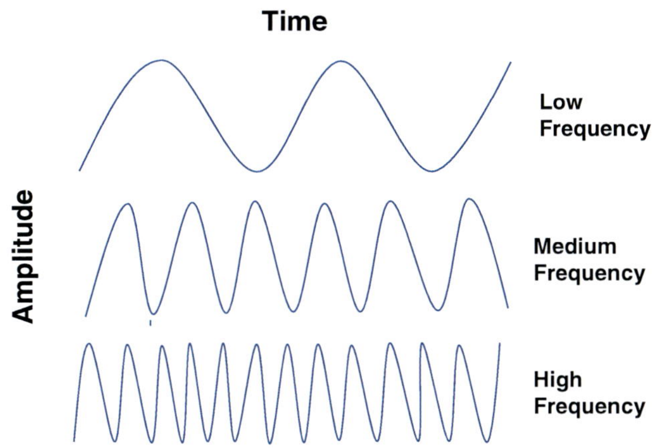

Figure 8.1 Example of frequency waves.

Table 8.1 Number of unreported leaks per km at average 40 m pressure 2001–2011.

	KM	Leaks	Leaks/Km	Leaks/100 Km
North America	1259	1056	0.84	83.88
Europe	743	924	1.24	124.36
Asia	158	94	0.59	59.49
Middle East	24	6	0.25	25.00
Totals	2184	2080	0.95	95.24

Table 8.2 Number of unreported leaks over mains diameter 2001–2011.

Mains Dia mm	KM	Leaks	Leaks/Km	Leaks/100 Km
<500	1139	1547	1.36	135.82
500–1000	987	515	0.52	52.18
>1000	58	18	0.31	31.03
Totals	2184	2080	0.95	95.24

8.4 LIFE OF A LEAK

Each leak has a life span and although there are technical terms available, as we are discussing the larger diameter mains the authors have at this point tried to explain them in simple terms.

There are 4 parts to the life of a leak

Weep – Leak – Burst – Catastrophic failure

Weep/Small loss – a small amount of water leaving a pipe from a small failure that is less than 5 litres per minute (1 imperial gallon per minute)

Leak/Medium loss – an amount of water that leaves a pipe through an orifice at an estimated flow of 90 litres per minute (20 imperial gallons per minute)

Burst/Large loss – an amount of water that leaves a pipe from an orifice at an estimated flow of 315 litres per minute (70 imperial gallons per minute)

Catastrophic failure – a complete rupture of the pipeline

In the initial request for data it was asked if any of the leaks excavated for repair had the flows measured by some method. It was reported that only some of the repairs had in fact had flows measured albeit some very crude. This data was used along with the average pressure reported as the base line to estimate the volume lost from a Weep to a Burst shown in Table 8.3.

Table 8.3 Estimated flows from unreported leaks at an average pressure of 40 m (From over 2000 incidents) all flows in imperial gallons.

Flow from a leak	Weep	Leak	Burst	Cat Failure
Gallons/hour	60	1200	4200	4200+
M^3/hour	0.27	5.4	19	19+

Note: leak flow lost increases with pressure

Further exercises have to take place to calibrate these figures as the method of measuring in many instances cannot be validated.

The numbers used in Table 8.3 have been rounded to match losses by imperial gallons.

It was then asked to categorise all other repaired leaks into the following section to give the number of leaks by type as used in Table 8.4.

- Weep <0.27 m^3/hr
- Leak 0.27–11 m^3/hr;
- Burst is 11 m^3/hr–27 m^3/hr
- Catastrophic failure is greater than 27 m^3/hr.

Table 8.4 Unreported leak categorisation.

Leak type	Weep	Leak	Burst	Cat Failure
Total Leaks Identified	143	1,278	659	0

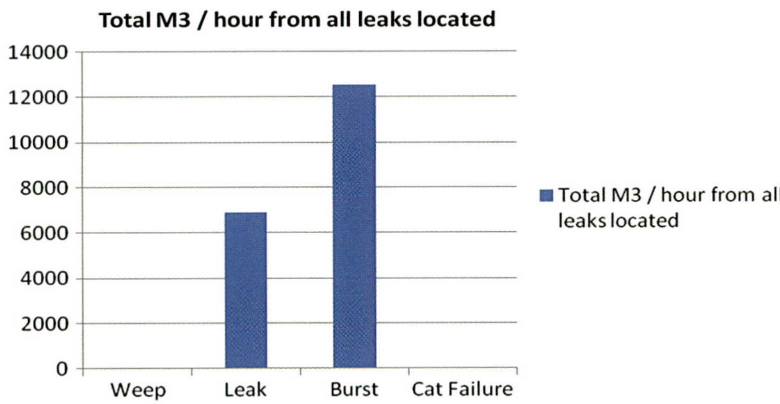

Note: To confirm all flows rates shown are based on a pressure of 40 m.

Figure 8.2 Volume lost from all leaks/hour/category.

8.5 Hz – LEAK NOISE

It is noticeable that there is a varied range in Hz when listening to a leak noise, this can be due to many factors ranging from pipe material, pipe diameter, pressure and leak size however it can be seen that when the internal hydrophone passes the location of the leak regardless of the material of the pipe line the leak noise is greater than 300 Hz and can easily be heard by the operator. This is due to the fact that the maximum distance the internal tethered hydrophone is away from the leak is that of the pipe diameter. The leak noise in Hz drops over a distance and starts to go beneath the human range, in some instances this occurs as quickly as 30 m from the leak position. It can be seen in the figures below that when using internal tethered devices the operator hearing ability is not critical but should be a consideration. Today's technology has software on the equipment which measures and reports these low frequency noises where as the operator may not be able to hear these.

Engineers have tried to use the frequency heard from a leak to estimate the size of the water leak. It is the authors' opinion that a leak cannot be measured as to the flow lost just by a noise measurement in Hz – there are too many other parameters which affect the range of losses.

The Figures 8.3–8.6 below show the Hz range observed over the distance measured in metres away from the various leaks.

The process used was to stop the internal hydrophone at pre-determined increments prior to the leak, at the leak and after the leak to measure the noise in Hz at each point; the figures were then built from this data.

It should be noted that the leak noise changes in the Hz range depending on the material and mains diameter, it should also be noted that in Figures 8.4 & 8.5 although the figures are the same leak the Hz range changes over the time the survey was conducted and this may have been due to a pressure variation over the period.

Viewing the Figures 8.3–8.6 they show the leak noise in Hz's when using internal an hydrophone drops below the threshold of human hearing within 40 m of the leak on metallic pipes and 25 m on MDPE pipes both with an average pressure of 35–41 m. Due to these observations it should be considered that leakage surveys are not completed on metallic mains diameter greater than 300 mm or non metallic mains regardless of diameter using traditional manual listening sticks.

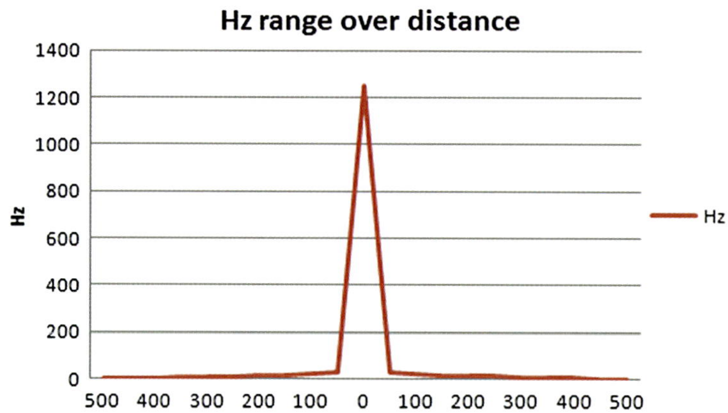

Figure 8.3 Noise in Hz 500 m from a leak on 300 mm Cast Iron pipe at 35 m pressure.

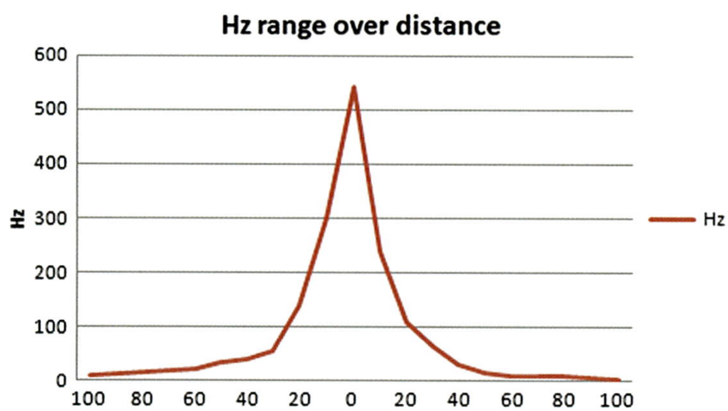

Figure 8.4 Noise in Hz 100 m from a leak on 500 mm Cast Iron pipe at 41 m pressure.

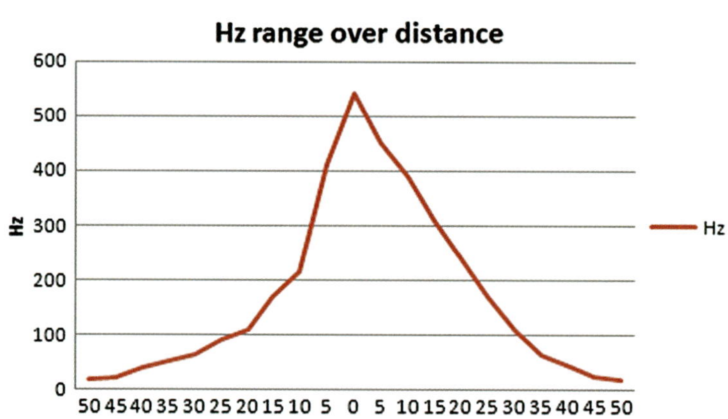

Figure 8.5 Noise in Hz 50 m from a leak on a 500 mm Cast Iron pipe at 41 m pressure.

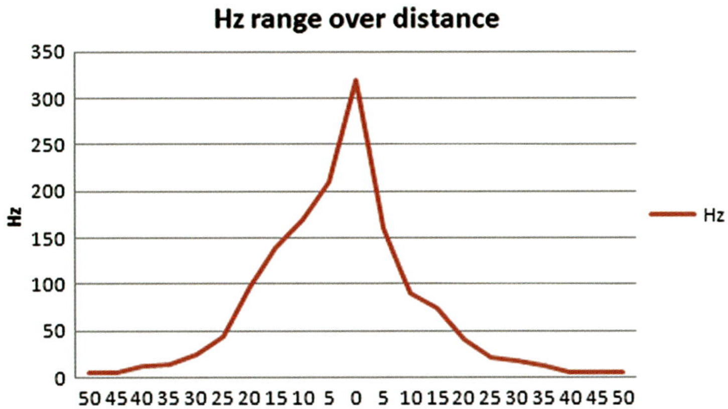

Figure 8.6 Noise in Hz 50 m from a leak on a 300 mm MDPE pipe 37 m pressure.

If using some form of electronic listening devices as the means to identify leak noises these should be able to identify if a leak is present below the 20 Hz range of noise, if they do not have this ability then we now know that leaks may be missed (Figure 8.7).

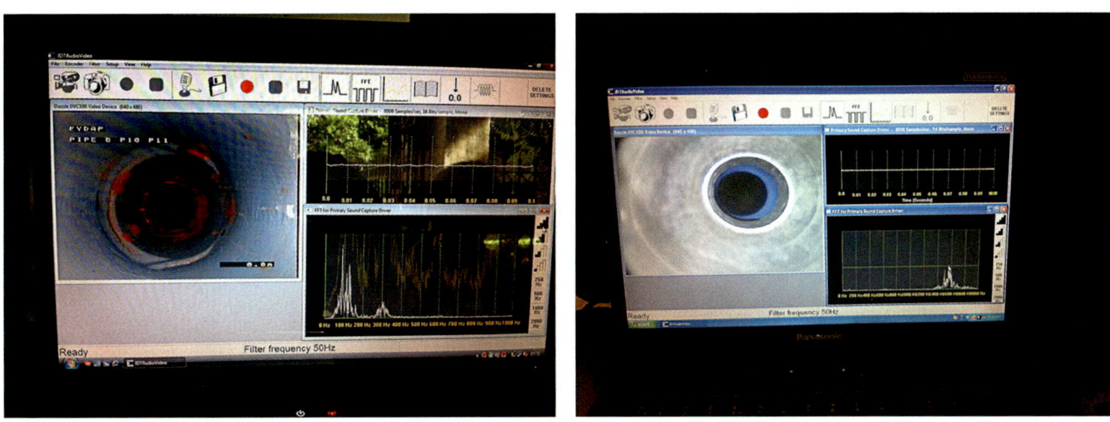

Figure 8.7 Screen dump from JD7 software showing of a leak inside a water pipe and the software showing the noise generated from the leak in Hz and Amplitude JD7AV software – version 1.1.0.10 developed 2012. JD7, Derby, UK. (*Source*: JD7 Ltd).

This conclusion falls inside the guide lines reported by the IWA WLSG Acoustic Initiative group and the written manual ‘Leakage Detection Methodologies including ALC Matrix’ Ver. 4 January 2011. Table 8.5 shows one of the 4 matrixes developed which show the best possible scenario and yet still does not promote manual listening on fittings for leakage surveys on any non metallic mains or metallic mains 300 mm and above.

It should also be noted that as a rule of thumb the noise frequency from a 100 mm non metallic main has the same acoustic parameters as that of a 400 mm metallic main.

Table 8.5 This matrix is for leakage detection on all property and mains fittings with pressures greater than 10 m head or 15 psi. Fittings are at a minimum distance of 10 m apart and maximum 50 m.

Diameter	mm	75	100	150	200	250	300	350	400	450	500	600	700	800	900	1000+
	inches	3	4	6	8	10	12	14	16	18	20	24	28	32	36	40+
Material																
Metallic all		A,B, C, D, F,G	A,B, C, D, F,G	A,B, C, D, F,G	A,B, C, D, F,G	A,B, C, D, F,G	A,C, D,E, F,G	A,C, D,E, F,G	A,C, D,E, F,G	C,D, E, F, G	C,D, E, F, G	C, D, E	C, D, E	D,E	D,E	D,E
Concrete all		A,C, D,F, G	A,C, D,F,G	A,C, D,F,G	A,D	A,D	A,D, E	A,D, E	A,D, E	E	E	E	E	E	E	E
Asbestos Cement		A,C, D,F, G	A,C, D,F,G	A,C, D,F,G	A,C, D	A,C, D	A,D, E	A,D, E	A,D, E	E	E	E	E	E	E	E
GRP		A,C, D,F, G	A,C, D,F,G	A,C, D,F,G	A,C, D	A,C, D	A,D, E	A,D, E	A,D, E	E	E	E	E	E	E	E
PVC		A,C, D,F, G	A,C, D,F,G	A,C, D,F,G	A,D	A,D	A,D, E	A,D, E	A,D, E	E	E	E	E	E	E	E
Polyethylene all		A,C, D,F, G	A,C, D,F,G	A,C, D,F,G	A,D	A,D	A,D, E	A,D, E	A,D, E	E	E	E	E	E	E	E

Method A Gas Injection
Method B Traditional Techniques with Manual Listening Stick
Method C Non-Intrusive Acoustic Techniques that is Standard Correlator, Correlating Noise Loggers (Accelerometers)
Method D Intrusive Acoustic Techniques that is Standard Correlator or Correlating Noise Loggers (Hydrophones)
Method E Inline Inspection Techniques (Tethered & Free-swimming)
Method F Noise Loggers (Non-Correlating), Non-Intrusive Magnetic Connection
Method G Electronic Amplified Listening Ground Microphone

8.6 CONCLUSIONS

Conducting a manual acoustic sounding exercise on metallic pipelines 300 mm diameter and above with fitments more than 100 m apart or on any non metallic pipe work using human hearing alone is a non effective way of conducting a survey. The Hz range from a leak does not travel to the fitting high enough for the human ear to be able to hear the leak. If an electronic device has the ability to measure noise in Hz then this may be an option. On previous surveys it has been reported that listening devices on non metallic pipes were not effective unless they were attached to access points that were close to the leak – roughly within 5 m (16 ft).

Operators using headsets attached to leak-noise correlators are on many occasions unable to hear leak sounds. The thought process currently with engineers is if no noise is heard, then no leak should exist. Operators of the correlators have been surprised to be able to locate leaks using the cross-correlation of leak sounds that they could not hear.

The authors recommend that manual sounding on fittings for leak detection purposes on non metallic mains or on metallic water mains over 300 mm in diameter should stop as the operators may not hear any leaks present and an alternative approach for this type of routine operation should be considered.

The higher Hz range from the leak is lost very quickly in the pipe wall and the water of the larger diameter pipes leaving the lower frequencies that can travel long distances – it is these lower frequencies that can be problematic to the traditional correlator. Insertion technology is successful since the acoustic head is only the maximum distance from the leak as the pipe diameter it is in that is on a 500 mm diameter pipe the acoustic sensor (hydrophone) inside the pipe when passing the leak can only be a maximum of 500 mm from the leak position.

Advancements in correlators are changing and some are beginning to show successes in locating leaks using the low frequencies over the longer distances.

The average number of leaks that have been found over the 2184 km is 95.24 per 100/km with varied results over the 4 regions the data was gathered from. Some reasons for these varied results can be the pressure within the water main. Within Europe where the average pressure is 45 m (67.5 psi) 124.3 leaks/100 km have been located where as in Asia where pressure have been as low as 3 m (5 psi) only 59.49 leaks/100 Km were found. This is slightly below what has been previously estimated at 110 leaks per 100/km and this may be due to the increase of surveys being completed in the Asia & Middle East regions. It has to be commented on that the results from Asia have been very varied with the number of leaks per km and this should be investigated to understand why the variation has occurred.

For ease the authors have divided the leaks into 4 categories with these ranging from a weep to a catastrophic failure. The numbers of leak identified within these categories were: Weeps 143, Leaks 1,278, Bursts 659 and Catastrophic Failures zero. The greatest number of leaks was found in the 'Leak' category however the greatest volume of water lost per hour was from the 'Burst' category. The number of Weeps located was found to save little water and are also the most expensive to repair, the average loss from the weep of 0.27 m^3/hr is close to the current limit for background leakage (0.25 m^3/hr at 50 m pressure), these events are not normally repaired if they are considered 'beyond economic repair and become equivalent to background leakage' but as the authors have mentioned these have to be investigated with as they will progress to a Leak. The time scale of this life has not been looked into and is unknown but with soil analysis and NDT surveys undertaken then this life expectancy can be calculated.

The findings from this paper were taken from companies throughout the world and are based on the information given at the time. The volume of water lost from these leaks has been estimated and further research has to be completed to give more accurate findings. The paper also at this time has commented on Hz range from leaks heard and further research has to be completed to identify more accurate results.

The authors would more than pleased to receive further information and/or data in this particular field in order to establish a firm base on which to build futute methodologies and strategies for dealing with leakage on large diameter mains.

ACKNOWLEDGEMENTS

Data supplied by JD7 Ltd and several water companies/consultants that wish to remain anonymous.

FURTHER READING

Hamilton S (2007). An Economic Active Leakage Control Policy without a Performance Indicator is not a Myth, IWA Water Loss 2007, 23rd–26th September 2007,Bucharest, Romania.

IWA Water Loss Specialist Group (2011). Leakage Detection Methodologies Manual Version, 4 January 2011.

Osama Hunaidi, Wing Chu, Alex Wang and Wei Guan (2000). Detecting Leaks in plastic pipes, *Journal AWWA*, **92**(2), 82–94.

Chapter 9

Paper 4: Technology – How far can we go?†

*Stuart Hamilton** and Daniel Krywyj***

Hydro Tec Ltd, The Barn, Thorpe Underwood, Northamptonshire, NN6 9PA, UK, E-mail: shamilton@hydrotec.ltd.uk

***JD7 Ltd, Unit 8, Masons Business Park Nottingham Rd, Derby, DE21 6YZ, UK E-mail: Daniel.Krywyj@jd7.co.uk*

Keywords: Technology, Acoustic, Correlator, ALC, AMR, Pressure, Software

9.1 INTRODUCTION

The world's population is increasing at a tremendous rate, the world's renewable water resources are reducing rapidly, the gap between water supply and demand is widening with urbanisation and climate change making the gap even wider. This paper outlines the importance of technology and innovation, describes technologies available in assisting water utilities to save valuable quantities of water lost through leaky networks and highlights the way forward in developing appropriate technology.

To deal with such losses in an effective manner, particularly from networks in water scarce areas, water utility managers are increasingly turning to technology to reduce costs, increase efficiency and improve reliability. Companies that continuously invest in technology and innovation will see a positive return on investment in terms of improving daily operations and collection and analysis of network data for decision making and forward planning.

9.2 BACKGROUND

It is evident that times are changing and globalisation, information technology, innovation and sustainable development are part of our daily lives. Water utilities are considered to be extremely conservative especially when it comes to change. It has been the approach for many years that if something works do not change it. However, a number of water companies recognised the need to go forward towards a higher standard of performance.

†This paper was originally presented as a paper for Water Loss 2012, 26–29 February 2012, Manila, Philippines, Organised by the IWA Water Loss Specialist Group.

Many water utilities are using Technology to improve reliability and to reduce costs. The need for change is becoming more and more pressing in order to facilitate and improve daily operations and to collect data for timely analysis and decision making as well as forward planning.

An important aspect that is usually debated is whether there is a positive return on investment in technology. In my opinion, there is a positive return on technology investment for companies that concentrate on enhancing system performance and focus on technological developments. This has certainly being the case with the technological changes which have taken place over the last 20 years in many water utilities around the globe.

Now, is there a real need to continue developing technology to control water losses? Let us first have a look at some facts and figures:

- Currently there are 7 billion human beings on the earth and this is increasing by more than 2 people per second, 173.000 per day or 63 million per year, each extra life needs food, energy water and shelter. This rate of increase means that by the year 2050 the earth's population will be 9.5 billion human beings. Over the past 2000 years nature has controlled the earth's population. However, now with the impact of extending life by controlling diseases, the majority of today's teenagers will live until they become grandparents.
- The earth is called the blue planet because 70% of its surface area is covered by water. Only 2.5% of this water is fresh water of which just 1% is available for human consumption; the rest is locked up in glaciers and polar ice caps.
- Today more than 1 billion people lack access to safe drinking water and this will worsen.
- Predictions indicate that in the next 20 years as much as 50% of the world's population will live in areas of water stress.

Such chronic shortages are often a result of poor infrastructure, politics, poverty or simply living in an arid part of the world. For instance, one of the richest cities in the world benefits from heavy annual rainfall, it supplies 20 million inhabitants with water and yet every day at least 1 million people experience some sort of water shortage.

Therefore, it is imperative that problems of water shortage and scarcity are dealt with in an effective manner, especially in water stressed areas. Losses from water distribution networks are a major issue worldwide. It is known that water leakage varies from less than 10% of the water put into network systems in extremely good situations to greater than 60% leakage in extremely poor situations with the average NRW being 20%–30% of the system input volume.

9.3 WHAT IS TECHNOLOGY?

- Application of knowledge to the practical aims
- Changing and manipulating the way we think
- Includes the use of materials, tools and techniques to make life easier, more pleasant, increase productivity and to ensure that the problem that was impossible becomes possible.
- Technology began to influence human endeavor as soon as people began using tools.

It is evident that Technology has an important role to play in order to deal efficiently and effectively with these high volume losses. So, where do we start and what do we have available? Well the very short answer is that technology is not moving fast enough to deal with these problems.

Methodologies are continuously changing to enable the best results to be achieved in the reduction of water losses. It is therefore imperative to move away from stock markets and profit making and to invest in the research and development of technology. Water companies and equipment manufacturers must work together in an effort to push current knowledge boundaries and to come up with improved and new ideas in order to complement current methodologies and together to provide solutions to reduce losses (Figure 9.1).

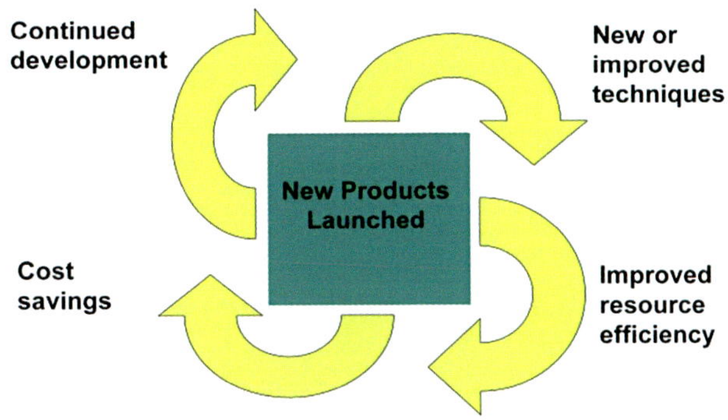

Figure 9.1 The continuous loop for product development.

Some countries do encourage research to be carried out through government funded projects but this is nowhere near enough to meet the ever increasing challenges that we are currently facing.

What is currently available and what is new to help combat the situation?

The IWA Water Loss Specialist Group identified the following methodologies which reduce real losses (leakages) as well as restrict apparent losses (increase revenues).

The methodologies for reducing Real Losses (leakages) are: Active Leakage Control, Pressure Management, Speed and Quality of Repairs and Renewal of Pipelines. Typical technology which is available in each one of these areas is briefly described below.

9.4 ACTIVE LEAKAGE CONTROL

Recent advances in equipment to help with location of leaks include correlating noise loggers, digital leak noise correlators with three sensors for better leak positioning and a ground microphone that can prioritise leaks by size and internal acoustic sensors for large diameter pipelines.

Internal tethered pipe technology is such that a cable can be inserted through a fire hydrant which has a head attached to the end. Within this head is a camera, hydrophone and location sonde. This system can go with or against the flow and has a maximum current range of 100 m, the ability here is to see the leak, hear the leak and to locate the head to pinpoint the position of the area of interest. This technology on the 300 mm and smaller metallic mains does not replace the traditional correlator however on the non metallic mains it has distinct advantages. This technology is used for internal condition pipe assessment showing tuberculation, unknown laterals, blockages and many other conditions. There is also a long range tethered technology available which can go distances currently up to 2000 m also the head combines a camera, hydrophone and location sonde. A company is in the process of attaching an ultra sonic internal condition assessment and pipe wall thickness measurement tool to the same head so it becomes a quadruple head.

Non tethered devices range from a ball that rolls down the main locating leaks as it passes them, this device is captured by a net and retrieved from the main, this technology is restricted to 300 mm diameter pipe work and above. A completely new device available to the market in 2012 is a small 3 cm × 10 cm bullet shaped device which can be launched in a 75 mm and above diameter pipe work and contains a camera and hydrophone. This device can be launched through a fitting on the main or a through bore fire hydrant. This device can be retrieved if it is attached to a tethered braid or can be launched as a free swimming device and captured somewhere down stream in the water main.

A robot is being designed and will be in commission during the next 24 months and this will live in the water main reporting on leaks as they occur and may have the ability to repair the water leak from inside the main with some sort of sealant it carries with it.

Other non acoustic technologies to assist in the location of leaks in the distribution system are being developed. One of these is software to localise leak positions using pressure drops and pressure variants in the network to locate the leak positions. Another uses statistical analysis of past data to try to calculate when and where a leak will occur next.

Another new technology is a system that measures flows through either permanent or temporary flow meters in the distribution system to indentify the area where an increase in flow has occurred, indicating a potential leak. Acoustic devices can be attached to the meter to help locate the leak position.

There are several European Union co – funded research projects on technology currently in progress – with varying results and successes. Two such projects that have now been completed are *'Waterpipe'* and *'Leaking'*. The projects have similar objectives – to provide a non-intrusive leak location technology.

9.5 WATERPIPE

'Waterpipe' is a *system* where the leak is located by ground penetrating imaging radar. The objectives of the project were to investigate and develop a high resolution imaging ground penetrating radar for the detection of pipes, leaks and damages to underground infrastructure - and to provide imaging of the damaged region.

A further enhancement was to produce an integrated system that will contain both the GPIR equipment and a Decision-Support-System (DSS) for the rehabilitation management of the underground water pipelines. This would use input from the inspections to assess, probabilistically, the time-dependent leakage and structural reliability of the pipelines and a risk-based methodology for rehabilitation decisions that considers the overall risk, financial, social and environmental criteria.

Please see Bimpas *et al.* (2010) for the findings from 'Waterpipe' project.

9.6 LEAKING

'Leaking' had objectives to investigate and develop an innovative leak inspection equipment for water pipelines based on microwave technology (a Continuous Wave Doppler radar, a Frequency Modulated Continuous Wave radar and a radiometer), and a decision support system, that stores available data on the pipe network, and receives input from leak inspections. It should be able to perform condition assessment to determine residual life time of the pipeline in question.

There are many other internationally funded projects – all of them are trying to achieve the breakthrough that would change the way Active Leakage Control is currently carried out. These initiatives are funded by Governments, universities, manufacturers, partnerships and water companies.

The author considers that not enough initiatives are undertaken or sufficient emphasis is given to reach a breakthrough which eventually will happen but unfortunately seems not to be on the horizon at the moment but with continuous research and development it could happen at any time.

The findings from 'Leaking' project will be available some time in 2010.

9.7 PRESSURE MANAGEMENT

The use of pressure reducing valves is always investigated with regard to obtaining the best possible results. Current pressure control techniques available are Flow/Time Modulation, Multi Point Control (flow or pressure), Critical Point Control (real time or through self learning algorithms) all of which provide solutions to problems of excess or varying pressure thus reducing losses.

New ideas in this field are available, one of which is using advanced programming to regulate the pressure valve thus always maintaining the required pressure at the control points, thus saving water above that of the traditional pressure valve operation.

9.8 SPEED AND QUALITY OF REPAIRS

An idea that is currently being investigated is that of self repairing pipes or self healing pipes. The pipe repairs itself from the inside using small particles or chemicals that are induced into the pipeline. Other similar ideas, such as when a leak occurs on the water main to send a report to a control station notifying the water company of the leak and its position are also looked into.

At the moment there are only a handful of cities in the world which have invested in such equipment, other cities failing to do so normally use the high investment cost as the reason.

9.9 RENEWAL OF PIPELINES

A perfect solution to leakage would be a pipe that doesn't leak. The pipes used today are designed to last 50+ years and be leak free; however, there is a major problem in that any joint that requires any sort of human intervention is unfortunately a source of a potential leak.

The question is often asked: 'Why hasn't the industry developed a pipe that is leak-free'? Many companies are in fact addressing this question - pipelines that are better protected against corrosion and at keeping the water clean.

Other investigations deal with the insertion of sensors either constructed in the pipe or alongside the pipe others are where a fibre optic cable is laid alongside the pipe reporting on changes to noise/temperature conditions all of which report back to a central computer when a leak occurs. This technology has to be introduced during the renewal of the pipeline, the major issue being that investment will be made in a system to locate water leaks on a pipe line that is designed to last for 50+ years. But this should not be the reason to reject such ideas - if the pipe fails for any reason prior to the end of its life of the pipe this will also be identified.

9.10 METHODOLOGIES IN REDUCING APPARENT LOSSES

Methodologies in reducing Apparent Losses thus increasing water utility revenues have been developed. Technologies to accurately measure water consumption and to reduce under-registration have recently been introduced and provide cost effective solutions.

9.11 METER ERROR – METER UNDER REGISTRATION

The under registration of flows through a water meter is a source of revenue lost by the utility and many companies have developed and produced devices that allow very small flows to be recorded, meters that measure low flows and meters that have no moving parts. Other devices are available that can be added to an existing meter to allow for extremely low flows to be measured.

9.12 AUTOMATIC METER READING

AMR is a way for the utility to read customer meters on a daily basis, sometimes continuously, allowing revenue to be collected using more frequent billing cycles. The company can also have a clear picture of the consumption by the household and indentify when a meter stops. This could indicate that the meter is broken or, for example, that a sole occupier is unwell. Conversely a higher than normal reading could indicate a leak on the service pipe.

Current technology allows transfer of data from the water meter by 'fixed radio network' or while a meter reader walks or drives past the meter. Software is available to produce a bill which is delivered by the utility employee while at the property.

One manufacturer has recently produced a drone aircraft that allows meters to be downloaded in remote locations by flying over the water meter. Noise loggers combined with water meters to allow both the meter flows and potential leaks being downloaded daily is another recent introduction.

9.13 SOFTWARE

In the area of water distribution network management advanced communication systems and software applications are playing an extremely important role in being able to take informed decisions timely and accurately.

9.14 COMMUNICATION SYSTEMS

The current trend is to apply solutions which combine information technology and telecommunications networks using the World Wide Web or GSM networks for the transfer of data which is obtained from site devices, such as water meters, pressures sensors, and so on.

Careful consideration and examination of the available technologies must be given in order to adopt an appropriate system with low capital expenditure as well as low operational and maintenance cost. A typical communication system for the storage and transfer of valuable information is shown diagrammatically in Figure 9.2 providing all the necessary information for the efficient and effective management of a water supply system.

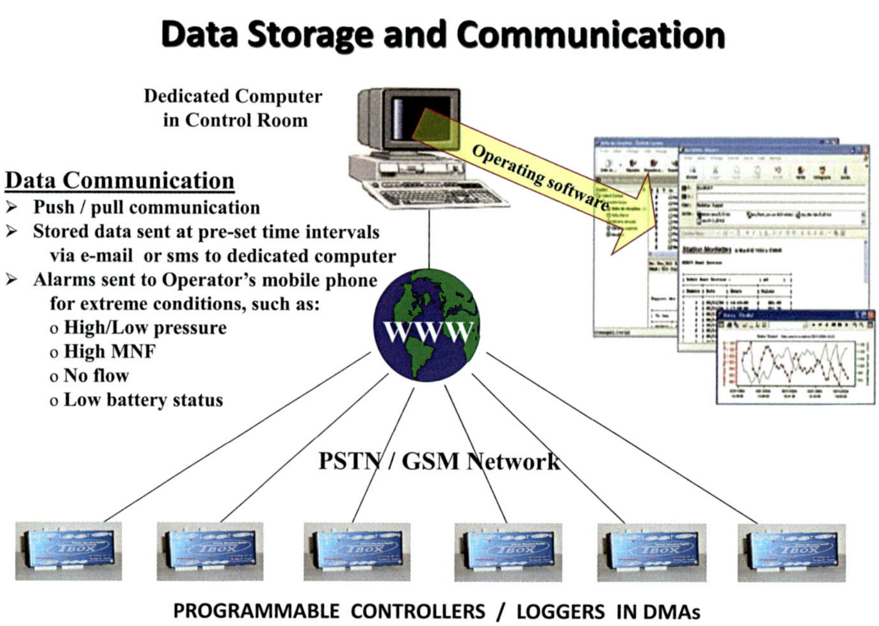

Figure 9.2 Typical set up for transfer of data collected and stored on critical site locations.

9.15 SOFTWARE APPLICATIONS

Operating software provides an intelligent communication interface between the monitoring stations and the central control which exploit the power of the internet by receiving data from the monitoring stations and has the capability of transferring and exporting data to the majority of data bases. Most software incorporates powerful graphics to display data in the form of graphs and statistical tables as well as to Geographical Information Systems for further analysis.

The pressing need to efficiently manage water distribution networks has highlighted the need to develop software tools that would assist in the integrated and automated management of the networks. Such asset management tools should assist the network owners to evaluate the condition of the water distribution network, assess historical incident data (leakage or breakage) and risk of failure, visualise areas of high risk, propose "repair or replace" strategies and prioritise the work based on the inherent risk and cost of action.

The risk assessment and management ("repair or replace") system is based on analytical and numerical modelling techniques and supplemented with geographical distribution systems (GIS). The goal is to enable water utilities to better manage condition assessment information, to process historical records with a number of analytical and numerical models, to identify underlying data patterns with artificial intelligence techniques and eventually to assess the corresponding risk of failure of each network element and to visually disseminate this information via geographical information systems.

9.16 INNOVATION IN THE FUTURE – CONCLUSIONS

The prime drivers needed for a continuous programme of new product development are shown diagrammatically in Figure 9.3. There is a market demand for new and improved technologies at affordable prices, but innovation and new products must be capable of delivering results in a cost efficient way. For this to happen a joint effort is needed by utilities, manufacturers, and researchers to develop the next generation of technology for the water utilities.

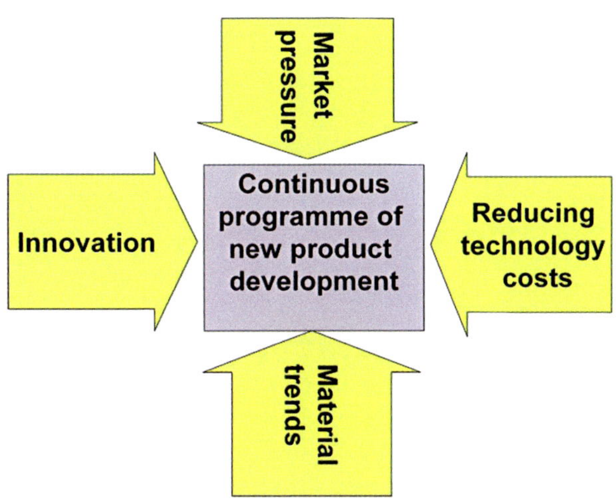

Figure 9.3 The Four drivers on the manufacturer.

It is felt by the author that governments should encourage such investigations using grants and that water companies should be willing to partner with manufacturers to design the solution to the water loss problem. The water scarcity issue is getting no better and we cannot afford to wait much longer to start to look for solutions.

It is illuminating to consider how far technology has moved forward in the past 10 years in one particular field – mobile telephones. Nowadays there are options for sound, vision, SMS, MMS, E-mail, internet, camera and radio – all of these in a small hand held device. These enhancements were developed to meet a market demand, but were also fuelled by competition. It is this competition and forward driven thinking that is required in the water industry.

All stakeholders should be willing to invest in solutions today – 'thinking outside the box' is an apt expression for moving forward with innovative technology for saving precious water in today's water scarce climate.

If we choose to cooperate can our intelligence and technology save us? It has to and we should be willing today to invest in the solution not wait for others to develop it tomorrow.

FURTHER READING

Bimpas M., Amditis A. and Uzunoglu N. (2010). Integrated High Resolution Imaging Radar and Decision Support System for the Rehabilitation of WATER PIPElines. IWA Publishing, London.

Kiss G., Koncz K. and Melinte C. (2007). Water pipe project – an innovative high resolution ground penetrating radar (GPIR) for detecting water leaks and a decision support system (DSS) for the rehabilitation management of pipelines. *Water Loss 2007 Conference Proceedings*, Bucharest, Romania, 23rd–26th September, Volume **3**, 621–631.

Hamilton S. (2005). Summary of conference proceedings. IWA 'Leakage 2005' conference, Halifax NV Canada, 12–14 September 2005.

Hamilton S. (2007a). An economic active leakage control policy without a performance indicator is not a myth. *Water Loss 2007 Conference Proceedings*, Bucharest, Romania, 23rd–26th September, Volume **3**, 752–762.

Hamilton S. (2007b). Acoustic principals in water loss management. *Water 21*, **9**(5), 47–48.

Hamilton S. and Hartley D. (2008). Misconceptions around acoustic leakage detection. *Water 21*, **10**(4), 54–56.

Hamilton S., Charalambous B. and Farley M. (2010). Technology: an integral part of water loss reduction. *Water 21*, **12**(1), 44–45.